Federal Emergency Management Agency
FEMA Overview & Emergency Preparedness Reference

Contents

Chapter 1

FEMA & Public Emergency Preparedness Overview

1.1 Federal Emergency Management Agency

"FEMA" redirects here. For other uses, see FEMA (disambiguation).

The **Federal Emergency Management Agency (FEMA)** is an agency of the United States Department of Homeland Security, initially created by Presidential Reorganization Plan No. 3 of 1978 and implemented by two Executive Orders on April 1, 1979.[1][4] The agency's primary purpose is to coordinate the response to a disaster that has occurred in the United States and that overwhelms the resources of local and state authorities. The governor of the state in which the disaster occurs must declare a state of emergency and formally request from the president that FEMA and the federal government respond to the disaster. FEMA also provides these services for territories of the United States, such as Puerto Rico. The only exception to the state's gubernatorial declaration requirement occurs when an emergency and/or disaster takes place on federal property or to a federal asset, for example; the 1995 bombing of the Alfred P. Murrah Federal Building in Oklahoma City, Oklahoma, or the Space Shuttle Columbia in the 2003 return-flight disaster.

While on-the-ground support of disaster recovery efforts is a major part of FEMA's charter, the agency provides state and local governments with experts in specialized fields and funding for rebuilding efforts and relief funds for infrastructure by directing individuals to access low interest loans, in conjunction with the Small Business Administration. In addition to this, FEMA provides funds for training of response personnel throughout the United States and its territories as part of the agency's preparedness effort.

1.1.1 History

Federal emergency management in the U.S. has existed in one form or another for over 200 years. FEMA's history is summarized as follows.

Prior to 1930s

A series of devastating fires struck the port city of Portsmouth, early in the 19th century. The 7th U.S. Congress passed a measure in 1803 that provided relief for Portsmouth merchants by extending the time they had for remitting tariffs on imported goods. This is widely considered the first piece of legislation passed by the federal government that provided relief after a disaster.[5]

Between 1803 and 1930, *ad hoc* legislation was passed more than 100 times for relief or compensation after a disaster. Examples include the waiving of duties and tariffs to the merchants of New York City after the Great Fire of New York (1835). After President Abraham Lincoln's assassination at John T. Ford's Theatre, the 54th Congress passed legislation compensating those who were injured in the theater.

Piecemeal approach (1930s–1960s)

After the start of the Great Depression in 1929, President Herbert Hoover had commissioned the Reconstruction Finance Corporation in 1932.[6] The purpose of the RFC was to lend money to banks and institutions to stimulate economic activity. RFC was also responsible for dispensing federal dollars in the wake of a disaster. RFC can be considered the first organized federal disaster response agency.

The Bureau of Public Roads in 1934 was given authority to finance the reconstruction of highways and roads after a disaster. The Flood Control Act of 1944 also gave the U.S. Army Corps of Engineers authority over flood control and

1

irrigation projects and thus played a major role in disaster recovery from flooding.

Department of Housing and Urban Development (1973–1979)

Federal disaster relief and recovery was brought under the umbrella of the Department of Housing and Urban Development (HUD), in 1973 by Presidential Reorganization Plan No. 1 of 1973, and the Federal Disaster Assistance Administration was created as an organizational unit within the department. This agency would oversee disasters such as occurring until its incorporation into the FEMA, Federal Emergency Management Agency, created by Presidential Reorganization Plan No. 2 of 1978, and implemented by Executive Orders 12127 and 12148.

Prior to implementation of Reorganization Plan No. 3 of 1978 by E.O. 12127 and E.O. 12148, many government agencies were still involved in disaster relief; in some cases, more than 100 separate agencies might be jockeying for control and jurisdiction of a disaster.[7]

Over the years, Congress increasingly extended the range of covered categories for assistance, and several presidential executive orders did the same. By enacting these various forms of legislative direction, Congress established a category for annual budgetary amounts of assistance to victims of various types of hazards or disasters, it specified the qualifications, and then it established or delegated the responsibilities to various federal and non-federal agencies.[8]

In time, this expanded array of agencies themselves underwent reorganization. One of the first such federal agencies was the Federal Civil Defense Administration, which operated within the Executive Office of the President. Functions to administer disaster relief were then given to the President himself, who delegated to the Housing and Home Finance Administration. Subsequently, a new office of the Office of Defense Mobilization was created. Then, the new Office of Defense and Civilian Mobilization, managed by the EOP; after that, the Office of Civil and Defense Mobilization, which renamed the former agency; then, the Office of Civil Defense, under the Department of Defense (DoD); the Department of Health, Education and Welfare (HEW); the Department of Agriculture; the Office of Emergency Planning (OEmP); the Defense Civil Preparedness Agency (replacing the OCD in the DoD); the Department of Housing and Urban Development (HUD) and the General Services Administration (GSA) (upon termination of the OEmP).[8]

These actions demonstrated that, during those years, the nation's domestic preparedness was addressed by several disparate legislative actions, motivated by policy and budgetary earmarking, and not by a single, unifying, compre-

hensive strategy to meet the nation's needs over time.[9] Then, in 1978 an effort was made to consolidate the several singular functions; FEMA was created to house civil defense and disaster preparedness under one roof. This was a very controversial decision.[8]

FEMA as an independent agency (1979–2003)

"National Fire Prevention and Control Administration" redirects here.

FEMA was established under the 1978 *Reorganization*

The FEMA seal before 2003

The FEMA flag before 2003

Plan No. 3, and activated April 1, 1979, by President Jimmy Carter in his Executive Order 12127.

In July, Carter signed Executive Order 12148 shifting disaster relief efforts to the new federal-level agency. FEMA absorbed the Federal Insurance Administration, the National

Fire Prevention and Control Administration, the National Weather Service Community Preparedness Program, the Federal Preparedness Agency of the General Services Administration and the Federal Disaster Assistance Administration activities from HUD. FEMA was also given the responsibility for overseeing the nation's Civil Defense, a function which had previously been performed by the Department of Defense's Defense Civil Preparedness Agency.

One of the disasters FEMA responded to was the dumping of toxic waste into Love Canal in Niagara Falls, New York, in the late 1970s. FEMA also responded to the Three Mile Island nuclear accident where the nuclear generating station suffered a partial core meltdown. These disasters, while showing the agency could function properly, also uncovered some inefficiencies.

In 1993, President Bill Clinton appointed James Lee Witt as FEMA Director. In 1996, the agency was elevated to cabinet rank.[10] This was not continued by President George W. Bush.[11] Witt initiated reforms that would help to streamline the disaster recovery and mitigation process. The end of the Cold War also allowed the agency's resources to be turned away from civil defense to natural disaster preparedness.[7]

After FEMA's creation through reorganization and executive orders, Congress continued to expand FEMA's authority by assigning responsibilities to it. Those responsibilities include dam safety under the National Dam Safety Program Act; disaster assistance under the Robert T. Stafford Disaster Relief and Emergency Assistance Act; earthquake hazards reduction under the Earthquake Hazards Reduction Act of 1977 and further expanded by Executive Order 12699, regarding safety requirements for federal buildings and Executive Order 12941, concerning the need for cost estimates to seismically retrofit federal buildings; emergency food and shelter under the Stewart B. McKinney Homeless Assistance Act of 1987; hazardous materials, under the Emergency Planning and Community Right-to-Know Act of 1986;

In addition, FEMA received authority for counter terrorism through the Nunn-Lugar-Domenici amendment under the Weapons of Mass Destruction Act of 1996, which was a response to the recognized vulnerabilities of the U.S. after the sarin gas attack on the Tokyo subway in 1995.[9]

Congress funded FEMA through a combination of regular appropriations and emergency funding in response to events.[12]

President George W. Bush signs the Homeland Security Appropriations Act of 2004

FEMA under Department of Homeland Security (2003–present)

Following the September 11, 2001, attacks, Congress passed the Homeland Security Act of 2002, which created the Department of Homeland Security (DHS) to better coordinate among the different federal agencies that deal with law enforcement, disaster preparedness and recovery, border protection and civil defense. FEMA was absorbed into DHS effective March 1, 2003. As a result, FEMA became part of the Emergency Preparedness and Response Directorate of Department of Homeland Security, employing more than 2,600 full-time employees. It became the Federal Emergency Management Agency again on March 31, 2007, but remained in DHS.

President Bush appointed Michael D. Brown as FEMA's director in January 2003. Brown warned in September 2003 that FEMA's absorption into DHS would make a mockery of FEMA's new motto, "A Nation Prepared", and would "fundamentally sever FEMA from its core functions", "shatter agency morale" and "break longstanding, effective and tested relationships with states and first responder stakeholders". The inevitable result of the reorganization of 2003, warned Brown, would be "an ineffective and uncoordinated response" to a terrorist attack or a natural disaster.[13]

Hurricane Katrina in 2005 demonstrated that the vision of further unification of functions and another reorganization could not address the problems FEMA had previously faced. The "Final Report of the Select Bipartisan Committee to Investigate the Preparation for and Response to Hurricane Katrina", released February 15, 2006, by the U.S. Government Printing Office, revealed that federal funding to states for "all hazards" disaster preparedness needs was not awarded unless the local agencies made the purposes for the funding a "just terrorism" function.[14]

Emergency management professionals testified that funds for preparedness for natural hazards was given less priority than preparations for counter terrorism measures. Testimony also expressed the opinion that the mission to mitigate vulnerability and prepare for natural hazard disasters before they occurred had been separated from disaster preparedness functions, making the nation more vulnerable to known hazards, like hurricanes.[15]

1.1.2 Organization

During the debate of the Homeland Security Act of 2002, some called for FEMA to remain as an independent agency. Later, following the failed response to Hurricane Katrina, critics called for FEMA to be removed from the Department of Homeland Security.[16] Today FEMA exists as a major agency of the Department of Homeland Security. The Administrator for Federal Emergency Management reports directly to the Secretary of Homeland Security. In March 2003, FEMA joined 22 other federal agencies, programs and offices in becoming the Department of Homeland Security. The new department, headed by Secretary Tom Ridge, brought a coordinated approach to national security from emergencies and disasters - both natural and man-made.

FEMA manages the National Flood Insurance Program. Other programs FEMA previously administered have since been internalized or shifted under direct DHS control.

FEMA is also home to the National Continuity Programs Directorate (formerly the Office of National Security Coordination). ONSC was responsible for developing, exercising, and validating agency wide continuity of operations and continuity of government plans as well as overseeing and maintaining continuity readiness including the Mount Weather Emergency Operations Center. ONSC also coordinated the continuity efforts of other Federal Executive Agencies.

FEMA began administering the Center for Domestic Preparedness in 2007.

Regions

- Regional Map

    - Region I, Boston, MA Serving: CT, MA, ME, NH, RI, VT

    - Region II, New York, NY Serving: NJ, NY, PR, USVI

    - Region III, Philadelphia, PA Serving: DC, DE, MD, PA, VA, WV

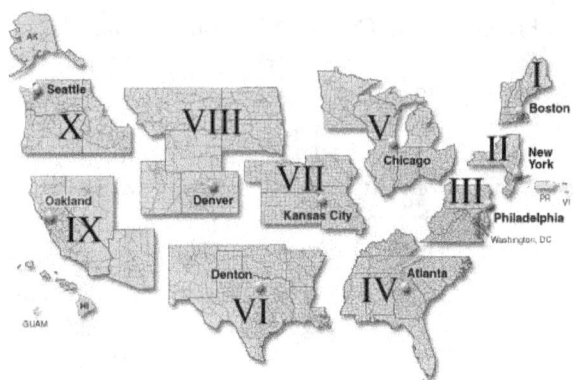

FEMA regions

- Region IV, Atlanta, GA Serving: AL, FL, GA, KY, MS, NC, SC, TN

- Region V, Chicago, IL Serving: IL, IN, MI, MN, OH, WI

- Region VI, Denton, TX Serving: AR, LA, NM, OK, TX

- Region VII, Kansas City, MO Serving: IA, KS, MO, NE

- Region VIII, Denver, CO Serving: CO, MT, ND, SD, UT, WY

- Region IX, Oakland, CA Serving: AZ, CA, HI, NV, GU, AS, CNMI, RMI, FM

- Region IX, PAO Serving: American Samoa, CNMI,Guam, Hawaii

- Region X, Bothell, WA Serving: AK (Alaska), ID, OR, WA

[17]

1.1.3 Pre-disaster mitigation programs

FEMA's Mitigation Directorate[18] is responsible for programs that take action before a disaster, in order to identify risks and reduce injuries, loss of property, and recovery time.[19] The agency has major analysis programs for floods, hurricanes, dams, and earthquakes.[19][20]

FEMA works to ensure affordable flood insurance is available to homeowners in flood plains, through the National Flood Insurance Program, and also works to enforce no-build zones in known flood plains and relocate or elevate some at-risk structures.[21]

Pre-Disaster Mitigation grants are available to acquire property for conversion to open space, retrofit existing buildings, construct tornado and storm shelters, manage vegetation for erosion and fire control, and small flood control projects.[22]

Taking Shelter From the Storm

The safe room construction plans and specifications from FEMA P-320, *Taking Shelter From the Storm*, are available in pdf and dwg format.[23]

1.1.4 Response capabilities

FEMA's emergency response is based on small, decentralized teams trained in such areas as the National Disaster Medical System (NDMS), Urban Search and Rescue (USAR), Disaster Mortuary Operations Response Team (DMORT), Disaster Medical Assistance Team (DMAT), and Mobile Emergency Response Support (MERS).

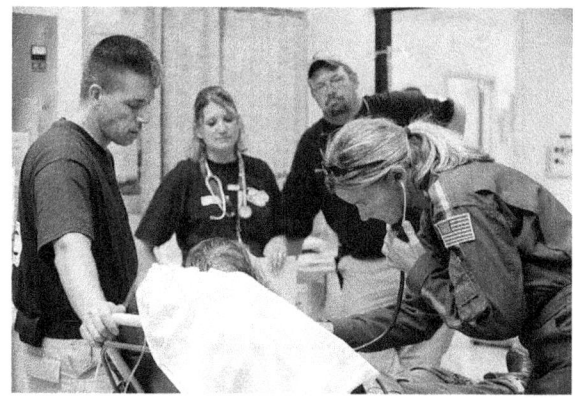

DMAT team deployed for Hurricane Ike in Texas

National Response Coordination Center (NRCC)

FEMA's National Response Coordination Center (NRCC) is a multiagency center located at FEMA HQ that coordinates the overall Federal support for major disasters and emergencies, including catastrophic incidents in support of operations at the regional-level. The FEMA Administrator,[24] or his or her delegate, activates the NRCC in anticipation of, or in response to, an incident by activating the NRCC staff, which includes FEMA personnel, the appropriate Emergency Support Functions, and other appropriate personnel (including nongovernmental organization and private sector representatives). During the initial stages of a response FEMA will, as part of the whole community, focus on projected, potential, or escalating critical incident activities. The NRCC coordinates with the affected region(s) and provides needed resources and policy guidance in support of incident-level operations. The NRCC staff specifically provides emergency management coordination, planning, resource deployment, and collects and disseminates incident information as it builds and maintains situational awareness—all at the national-level.[25] FEMA maintains the NRCC as a functional component of the NOC for incident support operations.[26][27]

An example of NRCC activity is the coordination of emergency management activities that took place in connection with the 2013 Colorado floods.[28]

National Disaster Medical System (NDMS)

The NDMS was transferred from the Department of Homeland Security to the Department of Health and Human Services, under the Pandemic and All-Hazards Preparedness Act, signed by President George W. Bush, on December 18, 2006.

NDMS is made of teams that provide medical and al-

lied care to disaster victims. These teams include doctors, nurses, pharmacists, etc., and are typically sponsored by hospitals, public safety agencies or private organizations. Also, Rapid Deployment Force (RDF) teams, composed of officers of the Commissioned Corps of the United States Public Health Service, were developed to assist with the NDMS.

Disaster Medical Assistance Teams (DMAT) provide medical care at disasters and are typically made up of doctors and paramedics. There are also National Nursing Response Teams (NNRT), National Pharmacy Response Teams (NPRT) and Veterinary Medical Assistance Teams (VMAT). Disaster Mortuary Operational Response Teams (DMORT) provide mortuary and forensic services. National Medical Response Teams (NMRT) are equipped to decontaminate victims of chemical and biological agents.

Urban Search and Rescue (US&R)

The Urban Search and Rescue Task Forces perform rescue of victims from structural collapses, confined spaces, and other disasters, for example mine collapses and earthquakes.

Mobile Emergency Response Support (MERS)

These teams provide communications support to local public safety. For instance, they may operate a truck with satellite uplink, computers, telephone and power generation at a staging area near a disaster so that the responders can communicate with the outside world. There are also Mobile Air Transportable Telecommunications System (MATTS) assets which can be airlifted in. Also portable cellphone towers can be erected to allow local responders to access telephone systems.

FEMA vehicle provides communications support after a major hurricane.

Preparedness for nuclear incidents

On August 1, 2008, FEMA had released "Planning Guidance for Protection and Recovery Following Radiological Dispersal Device (RDD) and Improvised Nuclear Device (IND) incidents"[29] which indicate action guide in case of radiation contamination. This Notice is specified as action guide for Radiological Dispersal Device (RDD) and Improvised Nuclear Device (IND) involving high levels of radiation. According to the Federation of American Scientists, during the Cold War FEMA prepared assessments of the likely consequences of a full-scale Soviet nuclear attack on the United States for use in planning mitigation and recovery efforts.[30]

Training

FEMA offers a large number of training classes, either at its own centers, through programs at the state level, in cooperation with colleges and universities, or online. The latter are free classes available to anyone, although only those with U.S. residency or work eligibility can take the final examinations. More information is available on the FEMA website under the "Emergency Personnel" and "Training" subheadings. Other emergency response information for citizens is also available at its website.

FEMA runs the Incident Workforce Academy, a two-week emergency preparedness training program for FEMA employees. The first class of the academy graduated in early 2014.[31]

The Training and Education Division within FEMA's National Integration Center directly funds training for responders and provides guidance on training-related expenditures under FEMA's grant programs. Catalog available at TED Course Catalog. Information on designing effective training for first responders is available from the Training and Education Division at First Responder Training. Emergency managers and other interested members of the public can take independent study courses for certification at FEMA's online Emergency Management Institute.

Emergency Management Institute training and certifications   See also: Emergency Management Institute

EMI offers credentials and training opportunities for United States Citizens. Note that students do not have to be employed by FEMA or be a federal employee for some of the programs.[32]

EMI maintains a strategic partnerships with Frederick Community College. FCC has contracted with the Emergency Management Institute to provide college credit for the Independent Study Program (ISP). FCC offers 8 specialized Letters of Recognition, an Undergraduate Certificate, and an Associate of Applied Science degree in Emergency Management.[33]

1.1.5   FEMA Corps

FEMA Corps, who range in age from 18–24 years old, is a cadre dedicated to disaster response and recovery. It is a new partnership between The Corporation for National and Community Service's AmeriCorps NCCC and FEMA.[34] The Corps, described as a "dedicated, trained, and reliable disaster workforce" works full-time for 10 months on federal disaster response and recovery efforts. Over 150 members of the inaugural FEMA Corps class graduated in June, 2013, at the AmeriCorps NCCC campus in Vicksburg, Mississippi. The Corps work on teams of 8 to 12 people, and follow the traditional NCCC model of living together and traveling together. In addition to working with FEMA, corps members must perform AmeriCorps responsibilities such as Physical Training three times a week, National Days of Service, and Individual Service Projects in communities throughout the United States. The Corps receives $4.75 a day for food, and a living stipend of approximately $4,000 over 10 months. An education award is distributed to corps members who successfully serve 10 months of service, completing 1,700 total hours.[35]

Donations management

FEMA has led a Public-Private Partnership in creating a National Donations Management Program making it easier for corporations or individuals not previously engaged to make offers of free assistance to States and the Federal Government in times of disaster. The program is a partner-

FEMA Corps Pacific Region Blue Unit

ship among FEMA, relief agencies, corporations/corporate associations and participating state governments. The technical backbone of the program is an online technology solution called The Aidmatrix Network which is managed by the independent nonprofit organization, Aidmatrix.

1.1.6 Criticism

Hurricane Andrew – 1992

See also: Hurricane Andrew

In August 1992, Hurricane Andrew struck the Florida and Louisiana coasts with 165 mph (265 km/h) sustained winds. FEMA was widely criticized for its response to Andrew, summed up by the famous exclamation, "Where in the hell is the cavalry on this one?" by Kate Hale, emergency management director for Dade County, Florida. FEMA and the federal government at large were accused of not responding fast enough to house, feed and sustain the approximately 250,000 people left homeless in the affected areas. Within five days the federal government and neighboring states had dispatched 20,000 National Guard and active duty troops to South Dade County to set up temporary housing. This event and FEMA's performance was reviewed by the National Academy of Public Administration in its February 1993 report "Coping With Catastrophe" which identified several basic paradigms in Emergency Management and FEMA administration that were causes of the failed response.

FEMA had previously been criticized for its response to Hurricane Hugo, which hit South Carolina in September 1989, and many of the same issues that plagued the agency during Hurricane Andrew were also evident during the response to Hurricane Katrina in 2005.

Additionally, upon incorporation into DHS, FEMA was legally dissolved and a new Emergency Preparedness and Response Directorate was established in DHS to replace it. Following enactment of the Post Katrina Emergency Man-

agement Reform Act of 2006 FEMA was reestablished as an entity within DHS, on March 31, 2007.

Southern Florida Hurricanes – 2004

South Florida newspaper *Sun-Sentinel* has an extensive list of documented criticisms of FEMA during the four hurricanes that hit the region in 2004.[36] Some of the criticisms include:

- When Hurricane Frances hit South Florida on Labor Day weekend (over 100 miles north of Miami-Dade County), 9,800 Miami-Dade applicants were approved by FEMA for $21 million in storm claims for new furniture; clothes; thousands of new televisions, microwaves and refrigerators; cars; dental bills; and a funeral even though the Medical Examiner recorded no deaths from Frances. A U.S. Senate committee and the inspector general of the Department of Homeland Security found that FEMA inappropriately declared Miami-Dade county a disaster area and then awarded millions, often without verifying storm damage or a need for assistance.[37][38]

- FEMA used hurricane aid money to pay funeral expenses for at least 203 Floridians whose deaths were not caused by the 2004 Hurricanes, the state's coroners have concluded. Ten of the people whose funerals were paid for were not even in Florida at the time of their deaths.[39]

Hurricane Katrina – 2005

Evacuees taking shelter at the Astrodome in Houston, Texas

See also: Criticism of government response to Hurricane Katrina

FEMA received intense criticism for its response to the Hurricane Katrina disaster in August 2005. FEMA had pre-positioned response personnel in the Gulf Coast region. However, many could not render direct assistance and were only able to report on the dire situation along the Gulf Coast, especially from New Orleans. Within three days, a large contingent of National Guard and active duty troops were deployed to the region.

The enormous number of evacuees simply overwhelmed rescue personnel. The situation was compounded by flood waters in the city that hampered transportation and poor communication among the federal government, state and local entities. FEMA was widely criticized for what is seen as a slow initial response to the disaster and an inability to effectively manage, care for and move those trying to leave the city.

Then-FEMA Director Michael D. Brown was criticized personally for a slow response and an apparent disconnection with the situation. Michael Brown would eventually be relieved of command of the Katrina disaster and soon thereafter resigned.

Katrina was seen as the first major test of the nation's new disaster response plan under DHS. It is widely held that many things did not function as planned.

According to the U.S. House of Representatives Select Bipartisan Committee to Investigate the Preparation for and Response to Hurricane Katrina:[40]

- "The Secretary Department of Homeland Security should have designated the Principal Federal Official on Saturday, two days prior to landfall, from the roster of PFOs who had successfully completed the required training, unlike then FEMA Director Michael Brown. Considerable confusion was caused by the Secretary's PFO decisions."

- "DHS and FEMA lacked adequate trained and experienced staff for the Katrina response."

- "The readiness of FEMA's national emergency response teams was inadequate and reduced the effectiveness of the federal response."

- "Long-standing weaknesses and the magnitude of the disaster overwhelmed FEMA's ability to provide emergency shelter and temporary housing."

- "FEMA logistics and contracting systems did not support a targeted, massive, and sustained provision of commodities."

- "Before Katrina, FEMA suffered from a lack of sufficiently trained procurement professionals."

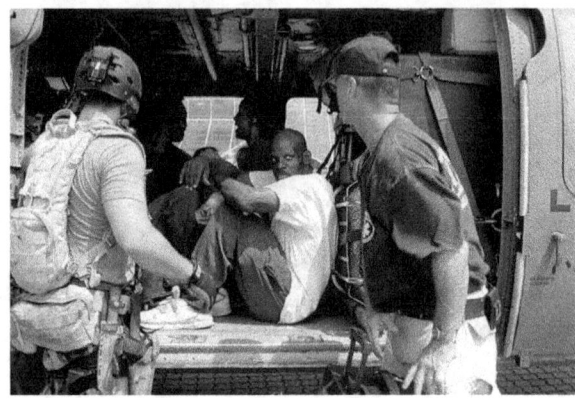

A DMAT member assures a rescued man that the trip to the airport will be safe.

Other failings were also noted. The Committee devoted an entire section of the report to listing the actions of FEMA.[41] Their conclusion was:

> For years emergency management professionals have been warning that FEMA's preparedness has eroded. Many believe this erosion is a result of the separation of the preparedness function from FEMA, the drain of long-term professional staff along with their institutional knowledge and expertise, and the inadequate readiness of FEMA's national emergency response teams. The combination of these staffing, training, and organizational structures made FEMA's inadequate performance in the face of a disaster the size of Katrina all but inevitable.[41]

Pursuant to a temporary restraining order issued by Hon. Stanwood R. Duval, United States District Court Judge, Eastern District of Louisiana as a result of the McWaters v. FEMA class-action, 7 February 2006 was set as the deadline for the official end of any further coverage of temporary housing costs for Katrina victims.[42][43]

After the February 7 deadline, Katrina victims were left to their own devices either to find permanent housing for the long term, or to continue in social welfare programs set up by other organizations. There were many Katrina evacuees living in temporary shelters and/or trailer parks set up by FEMA and other relief organizations in the first months after the disaster hit, but many more were still unable to find housing.

In July 2007, ice that had been ordered for Katrina victims but had never been used and kept in storage facilities, at a cost of $12.5 million, was melted down.[44]

In June 2008, a CNN investigation found that FEMA gave

away about $85 million in household goods meant for Hurricane Katrina victims to 16 other states.[45]

Buffalo snowstorm – 2006

FEMA came under attack for their response to the October Surprise Storm on 13 October 2006 in Buffalo, New York. As FEMA legally cannot interfere with state business unless asked, FEMA responded that as per procedure, the Governor of the state of New York had not asked for FEMA's assistance. FEMA Headquarters had been in constant contact with State congressional offices providing them with the latest information available. Claims state that FEMA officials did not arrive until 16 October, three days after the storm hit. The damage by this time included downed power wires, downed trees, and structural damage to homes and businesses.[46]

Dumas, Arkansas, tornadoes – 2007

Many people of Dumas, Arkansas, especially victims of the February 24, 2007 tornadoes, criticized FEMA's response in not supplying the amount of new trailers they needed, and only sending a set of used trailers, lower than the needed quantity. Following the storm, U.S Senator Mark Pryor had criticized FEMA's response to the recovery and cleanup efforts.[47]

California wildfires – 2007

FEMA came under intense criticism when it was revealed that a press conference on the October 2007 California wildfires was staged. Deputy Administrator Harvey E. Johnson was answering questions from FEMA employees who were posing as reporters. Many of these questions were "soft ball" questions (i.e., "Are you happy with FEMA's response so far?"), intentionally asked in a way that would evoke a positive response giving the impression that FEMA was doing everything right. In this way, any scrutiny from real reporters (many of whom were only given a 15-minute notice) would have been avoided. Fox News, MSNBC, and other media outlets aired the staged press briefing live.[48] Real reporters were notified only 15 minutes in advance and were only able to call in to a conference line, which was set to "listen-only" mode. The only people there were primarily FEMA public affairs employees.[49]

1.1.7 List of FEMA heads

As director of the Office of Emergency Preparedness

As director of FEMA (cabinet-level between 1996–2001)[10][11]

As Undersecretary of Emergency Preparedness and Response and Director of FEMA
(within the Department of Homeland Security)

As Undersecretary for Federal Emergency Management and Director of FEMA
(within the Department of Homeland Security)

As Administrator of the Federal Emergency Management Agency
(within the Department of Homeland Security)

On March 4, 2009, President Barack Obama nominated Florida's state emergency management director, W. Craig Fugate, to lead FEMA.

1.1.8 See also

- Civil defense by country

- Civil Contingencies Secretariat, United Kingdom equivalent emergency management agency

- Council of Governors

- FEMA camps conspiracy theory

- FEMA photo library

- FEMA trailer

- National Emergency Technology Guard

- PDD-62

- Emergency Preparedness Canada - Canadian counterpart disaster response agency

- U.S. Fire Administration

1.1.9 References

[1] "Executive Order 12127--Federal Emergency Management Agency". Federation of American Scientists.

[2] "About FEMA". Federal Emergency Management Agency. October 20, 2011.

[3] "FEMA's FY 2013 Budget Request". Federal Emergency Management Agency. April 13, 2013.

[4] Woolley, Lynn (September 12, 2005). "FEMA - Disaster of an Agency". Retrieved December 12, 2007. See Federation of American Scientists reference above for effective date of April 1, 1979, stated in Executive Order 12127.

[5] History of Federal Domestic Disaster Aid Before the Civil War, Biot Report #379: July 24, 2006. Suburban Emergency Management Project.

[6] Article on the RFC from EH.NET's Encyclopedia. Archived May 17, 2015 at the Wayback Machine

[7] "FEMA History". Federal Emergency Management Agency.

[8] Bea, Keith, "Proposed Transfer of FEMA to the Department of Homeland Security", Order Code RL31510 (updated 29 July 2002), Report for Congress, Congressional Research Service: Library of Congress.

[9] Falkenrath, Richard S., "Problems of Preparedness: U.S. Readiness for a Domestic Terrorist Attack" (2001)International Security, Boston.

[10] "President Clinton Raises FEMA Director to Cabinet Status" (Press release). Federal Emergency Management Agency. February 26, 1996. Retrieved March 3, 2010.

[11] Fowler, Daniel (November 19, 2008). "Emergency Managers Make It Official: They Want FEMA Out of DHS". CQ Politics. Retrieved March 3, 2010. During the Clinton administration, FEMA Administrator James Lee Witt met with the cabinet. His successor in the Bush administration, Joe M. Allbaugh, did not. (Archived by WebCite at Webcitation.org)

[12] Murry, Justin (updated July 10, 2006). "Emergency Supplemental Appropriations Legislation for Disaster Assistance: Summary Data FY1989 to FY2006", CRS Report for Congress, Congressional Research Service: The Library of Congress.

[13] Grunwald, Michael, and Susan B. Glasser (December 23, 2005). "Brown's Turf Wars Sapped FEMA's Strength". The Washington Post. p. A01. Retrieved April 18, 2007.

[14] Senate Bipartisan Committee (February 15, 2006), "The Final Report of the Select Bipartisan Committee to Investigate the Preparation for and Response to Hurricane Katrina, U.S. Government Printing Office: Washington, D.C.

[15] Senate Bipartisan Committee, 2006, p. 208.

[16] Serving America's Disaster Victims: FEMA Where Does it Fit? Homeland Security Policy Institute. January 13, 2009.

[17] http://www.fema.gov/regional-operations

[18] "Mitigation". Federal Emergency Management Agency. Archived July 1, 2012 at the Wayback Machine

[19] "FEMA's Mitigation Directorate Fact Sheet". Federal Emergency Management Agency.

[20] HAZUS is a computer model for hurricane, earthquake, and flood damage estimates. Archived July 4, 2012 at the Wayback Machine

[21] . Federal Emergency Management Agency.

[22] "Grant Program Comparison: Mitigation Division Grant Programs". Archived January 11, 2014 at the Wayback Machine

[23] FEMA Taking Shelter From the Storm

[24] http://www.fema.gov/media-library-data/20130726-1914-25045-1246/final_national_response_framework_20130501.pdf

[25] FEMA's State-of-the-Art National Response Coordination Center

[26] National Response Framework. May 2013. p. 43.

[27] National Response Coordination Center: It Takes A Whole Community for Response | FEMA.gov

[28] Homeland Security Today: FEMA Monitors Colorado Flooding; Supports State, Local Response

[29] FEMA, DHS "Planning Guidance for Protection and Recovery Following RDD and IND incidents" retrieved July 6, 2011.

[30] "Nuclear Attack Planning Base – 1990.

[31] Limardo, Jessica. "First FEMA Incident Workforce Academy class graduates". BioPrepWatch. February 13, 2014. (Retrieved 02-13-2014).

[32] EMI Program Info

[33] "Emergency Management". Frederick.edu. Frederick Community College. 4 April 2014. Archived from the original on 20 Feb 2014. Retrieved 20 April 2014.

[34] Announcing the Creation of FEMA Corps. FEMA.gov (2012-06-16). Retrieved on 2013-08-16.

[35] Welcome to the FEMA Corps Inaugural Class | Homeland Security. Dhs.gov (2012-09-14). Retrieved on 2013-08-16.

[36] "Sun-Sentinel Investigation: FEMA". Sun-Sentinel. Retrieved April 18, 2007.

[37] Kestin, Sally, and Megan O'Matz (October 10, 2004). "FEMA Gave $21 Million in Miami-Dade, Where Storms Were 'Like a Severe Thunderstorm'". Sun-Sentinel. Retrieved April 18, 2007.

[38] Kestin, Sally (June 8, 2005). "Homestead Women Sentenced to Probation for Cheating FEMA". Sun-Sentinel. Retrieved April 18, 2007.

[39] Kestin, Sally; Megan O'Matz; and Jon Burstein (August 10, 2005). "FEMA Paid for at Least 203 Funerals Not Related to 2004 Hurricanes". Sun-Sentinel. Retrieved April 18, 2007.

[40] "Executive Summary, Select Bipartisan Committee to Investigate the Preparation for and Response to Hurricane Katrina". February 15, 2006. U.S. Government Printing Office. Retrieved June 11, 2007. Archived September 2, 2013 at the Wayback Machine

[41] "FEMA, Select Bipartisan Committee to Investigate the Preparation for and Response to Hurricane Katrina". February 15, 2006. U.S. Government Printing Office. Retrieved June 11, 2007. Archived September 2, 2013 at the Wayback Machine

[42] Duval, Stanwood R., Jr.; United States District Court; Eastern District of Louisiana (December 12, 2005). ""Order of December 12, 2005" (Rec. Doc. No. 63)" (PDF). *"Beatrice B. Mcwaters, et al. v. Federal Emergency Management Section 'K' (3)" (No. 05-5488)*. USCourts.gov. Retrieved April 18, 2007.

[43] Duval, Stanwood R., Jr.; United States District Court; Eastern District of Louisiana. ""Modified Order of January 12, 2006" (Ref. Doc. No. 74)" (PDF). *"Beatrice B. Mcwaters, et al. v. Federal Emergency Management Section 'K' (3)" (No. 05-5488)*. USCourts.gov. Retrieved April 18, 2007.

[44] "FEMA To Melt Ice Stored Since Katrina". CBS News.

[45] FEMA Gives Away $85 Million of Supplies for Katrina Victims". CNN.

[46] "FEMA Replies to Unjustified Claims Regarding FEMA's Response To Early Snowstorm In Western New York". Federal Emergency Management Agency. Archived January 28, 2012 at the Wayback Machine

[47] "Ark. Pols Blast FEMA for Tornado Response". *USA Today*.

[48] "FEMA Stages Press Conference: Staff Pose As Journalists And Ask 'Softball' Questions"

[49] Ripley, Amanda (October 28, 2007). "Why FEMA Fakes It with the Press". *Time*.

1.1.10 Further reading

- MSNBC Article Senate panel recommends abolishing FEMA

- Federal Emergency Management: A Brief Introduction from the Congressional Research Service

1.1.11 External links

- Official website

- Federal Emergency Management Agency in the Federal Register

-

- EMI Emergency Management Higher Education Program

- FEMA Independent Study Program (ISP) Professional Development Series

- FEMA Photo Library

- Works by Federal Emergency Management Agency at Project Gutenberg

- Works by or about Federal Emergency Management Agency at Internet Archive

1.2 Public health emergency (United States)

In the United States, a **public health emergency** declaration releases resources meant to handle an actual or potential public health crisis. Recent examples include incidents of flooding, severe weather,[1] and the 2009 H1N1 influenza outbreak. Homeland Security Secretary Janet Napolitano described it as a "declaration of emergency preparedness."[2]

The National Disaster Medical System Federal Partners Memorandum of Agreement defines a public health emergency as *"an emergency need for health care [medical] services to respond to a disaster, significant outbreak of an infectious disease, bioterrorist attack or other significant or catastrophic event. For purposes of NDMS activation, a public health emergency may include but is not limited to, public health emergencies declared by the Secretary of HHS [Health and Human Services] under 42 U.S.C. 247d, or a declaration of a major disaster or emergency under the Robert T. Stafford Disaster Relief and Emergency Assistance Act (Stafford Act), 42 U.S.C. 5121-5206)."*[3][4]

The declaration of public health emergency in the March 2009 flood of the Red River in North Dakota was made under section 319 of the Public Health Service Act. Under section 1135 of the Social Security Act, this declaration permits the state government to request waivers of certain Medicare, Medicaid and CHIP requirements from the Centers for Medicare and Medicaid Services (CMS) Regional Office. Examples include allowing Medicare health plan beneficiaries to go out of network, allowing critical access hospitals to take more than the statutorily mandated limit of 25 patients, and not counting the expected longer lengths of stay for evacuated patients against the 96-hour average.[1]

In the swine flu outbreak, the declaration allowed the distribution of a federal stockpile of 12 million doses of Tamiflu to places where states could quickly get their share if they decide they need it, with priority going to the five states with known cases.[2] Because Obama's choice for Secretary of HHS, Kathleen Sebelius, had not yet been confirmed, the public announcement of the emergency was made by President Obama and Homeland Security Secretary Napolitano.[5] However, the formal determination of a

FEMA - 7797 - Photograph by Jocelyn Augustino taken on 03-12-2003 in District of Columbia

public health emergency was made by Charles Johnson, acting HHS secretary, under section 319 of the Public Health Service Act, 42 U.S.C. § 247d.[6]

A **military health emergency** is defined by the NDMS as *"an emergency need for hospital services to support the armed forces for casualty care arising from a major military operation, disaster, significant outbreak of an infectious disease, bioterrorist attack, or other significant or catastrophic event."*[3][4]

1.2.1   Catastrophe

A Catastrophe has been defined by Dr. Rick Bissell as: "an event that directly or indirectly affects and entire country, requires national and possibly international response, and threatens the welfare of a substantial number of people for an extended period of time."[7] Bissell makes the distinction between "catastrophes" and "disasters" in that, for catastrophes: "their complexity and their various impacts are so significant that the ordinary emergency planning, preparedness, and response tools are no longer sufficient, or may even be counterproductive."[7] Therefore, the failure of all levels of the disaster wheel (preparation, mitigation, response, recovery) can neither be planned for nor be fixed once it has occurred. Not all situations can be accounted for simply because of the combination of the common uncertainty and probability principles. It would be extremely lengthy due to the immense amount of detail needed to cover all aspects of any possible catastrophe would not only be daunting to read, even more so to develop realistically. The best way to begin to plan for a catastrophe would be by following the continuum of magnitude, starting at the local level and moving up all the way into international mutual aid. Because catastrophes have such a massive impact and demand resources, international aid should not be out of the question. The catastrophe plan should not be made

rigid because of the large amount of grey area as to what exactly will be affected by the event.

Prep Cycle

Public health officials could set up parameters to help prevent this event from occurring in the first place because they are able to send inspectors in a timely fashion to uphold the health codes. An outbreak of bubonic plague in this era may be a sign of biological attack. However, should such an outbreak occur again, the public health official would have to notify the Emergency Management at the first sign. Steps toward containment and recovery must be swift due to the nature of such an outbreak. Furthermore, there would need to be several tests taken and careful watch over possible target populations in order to quickly spot other new outbreaks. The origin of the outbreak must be determined quickly to prevent possible global spread. The emergency manager has access to the resources needed to contain and eliminate the outbreak, and combined with the education of the public health official on how to go about doing so would result in a better chance of containment. Also, the emergency manager would work on recovery steps for the affected population and infrastructure while the public health official addresses how to prevent an outbreak from occurring again. By using the preparedness cycle, planning transitions into organization, training and equipment, which in turn transitions to exercising those actions, followed by evaluating the effectiveness and improving upon the plan for next time.[[Sylves, Richard T. and William L. Waugh, Jr. (1996). Disaster Management in the U.S. and Canada: The Politics, Policymaking, Administration and Analysis of Emergency Management, 2nd edition. Springfield, IL: Charles C. Thomas, Publisher, LTD (ISBN 0-398-06609-4).]] Autonomy in terms of emergency management is a very important facet of the big picture. We want to be autonomous for most emergencies and disasters simply because locally, we know the populations, special needs populations, topography and key affected points much better and are able to respond faster and more effectively than outside organizations. However we as emergency managers do not want to be completely autonomous to all situations because in the case that we may actually need external help, we need to already have those paths carved out before hand to make logistics easier. Swift and effective response is key to optimal disaster recovery. As emergency managers we must be able to operate completely alone until mutual aid arrives. If our responses fail or if communications to other organizations fail, our efforts are wasted and the situation then becomes a possible catastrophe. We must be prepared for these events and so, education, training and communication are the three most important aspects. The Stafford Disaster Relief and Emergency Assistance Act and Posse Comitatus Act are just a couple of examples of steps taken to preserve local and state's right to autonomy from the federal and international governments.

1.2.2 Steps for response

Assess the upcoming event and the possible key affected areas. Figure the resources needed to do target hardening steps to help prevent total destruction and help with quick recovery. An old adage states that "an ounce of prevention is worth a pound of cure". Pre-planning stages and paving effective communication lanes to obtain external aid is very important to the saving of lives. Using these pre-made venues saves precious time during an event. A big obstacle to this is to make the happy medium between a very specific detailed emergency plan and a flexible one to which many disaster events may apply. Up-to-date education and training are other key elements to making for a swift and effective response and recovery. Retaining infrastructure is a huge challenge to an emergency manager because intact infrastructure makes for easier use of communication and multiple routes for much needed resourced to enter the affected area. Without this communication to the external organizations, they will not know how to bring about the needed resources (by land, air, sea?). The use of a command center away from the affected area such as a Louisiana emergency management center in Texas will help prevent this issue and many others by not "putting all of your eggs in one basket". If this chain of communication were to break down, the situation very well can turn from simple regional emergency to a catastrophe or even a mass extinction event.[8]

1.2.3 See also

- Public Health Emergency.gov

- Public Health Information Network

- Public Health Emergency Preparedness

- United States Public Health Service#Emergency response since 1999

- Surgeon General of the United States

- Public health laboratory

- United States Public Health Service

- United States Public Health Service Commissioned Corps

- Disaster Medical Assistance Team

- Medical Reserve Corps

- Office of the Assistant Secretary for Preparedness and Response

- United States Deputy Secretary of Health and Human Services

- Public Readiness and Emergency Preparedness Act

- Model State Emergency Health Powers Act

- Federal Emergency Management Agency

- Department of Health, Education and Welfare

- Emergency Management Institute

A number of state and local agencies, laws, and office holders have been omitted here.

1.2.4 References

[1] "HHS Acting Secretary Declares Public Health Emergency for North Dakota Storms". 2009-03-25.

[2] "US Declares Public Health Emergency for Swine Flu". Associated Press. 2009-04-26.

[3] "A Public Health Emergency from the Perspective of the U.S. National Disaster Medical System (NDMS)". 2007-04-10.

[4] "NATIONAL DISASTER MEDICAL SYSTEM MEMORANDUM OF AGREEMENT AMONG THE DEPARTMENTS OF HOMELAND SECURITY, HEALTH AND HUMAN SERVICES, VETERANS AFFAIRS, AND DEFENSE" (PDF). 2005-09-26.

[5] Carrie Budoff Brown (2009-04-26). "As flu hits, holes in W.H. health team". Politico.

[6] "Determination that a Public Health Emergency Exists". HHS.gov.

[7] Bissell, Rick. "Instructor Notes for Session No. 1". *Catastrophe Readiness and Response*. fema.gov. Retrieved 13 November 2012.

[8] Anderson, James (2011). *Public Policymaking*, 7th edition. Boston, MA: Wadsworth, Cengage Learning (ISBN 10: 0-618-97472-5)

- Sylves, Richard T. and William L. Waugh, Jr. (1996). Disaster Management in the U.S. and Canada: The Politics, Policymaking, Administration and Analysis of Emergency Management, 2nd edition. Springfield, IL: Charles C. Thomas, Publisher, LTD (ISBN 0-398-06609-4).

- Anderson, James (2011). Public Policymaking, 7th edition. Boston, MA: Wadsworth, Cengage Learning (ISBN 10: 0-618-97472-5).

1.3　Public Health Emergency Preparedness

In the United States government, the **Office of the Assistant Secretary for Preparedness and Response (ASPR)**, formerly the Office of Public Health Emergency Preparedness (or OPHEP), is a branch of the U.S. Department of Health and Human Services.

The Office of Public Health Emergency Preparedness, was established in June 2002 at the request of Tommy Thompson. In July 2006, a bill to amend the Public Health Service Act with respect to public health security and all-hazards preparedness and response was introduced. On December 19, 2006 it became public law and OPHEP was officially changed to the Office of the Assistant Secretary for Preparedness and Response.

Its scope of activity includes but is not limited to preparedness for bioterrorism, chemical and nuclear attack, mass evacuation and decontamination. [1] The ASPR the Secretary's principal advisor on matters related to bioterrorism and other public health emergencies. They are responsible for coordinating interagency activities between HHS, other Federal departments, agencies, offices and State and local officials responsible for emergency preparedness and the protection of the civilian population from acts of bioterrorism and other public health emergencies. The ASPR also works closely with global partners to address common threats around the world, enhancing national capacities to detect and respond to suchthreats, and to learn from each other's experiences as another step toward national health security for the United States and other countries.[2]

The first head of OPHEP was Donald Henderson, credited with having previously eradicated Smallpox. Soon Jerry Hauer, a veteran public health expert, took over as director, with Henderson taking a different role in the department. Hauer was removed from the job primarily for conflicts he had with Scooter Libby over whether the risks of smallpox vaccination were worth the benefit. Hauer charged that the Office of the Vice President was pushing for the universal vaccination despite the vaccine's health risks, primarily exaggerate the risk of biological terrorism.

RADM W. Craig Vanderwagen, M.D., was sworn into office on March 27, 2007 as the first Assistant Secretary for Preparedness and Response and recently retired.

RADM Nicole Lurie, MD, MSPH is the current ASPR.

The United States National Response Framework (NRF) is part of the National Strategy for Homeland Security that presents the guiding principles enabling all levels of domestic response partners to prepare for and provide a unified national response to disasters and emergencies. Building on the existing National Incident Management System (NIMS) as well as Incident Command System (ICS) standardization, the NRF's coordinating structures are always in effect for implementation at any level and at any time for local, state, and national emergency or disaster response.

1.3.1　References

[1] http://www.ahrq.gov/prep/cbrne/

[2] http://www.phe.gov/preparedness/international/pages/default.aspx

1.3.2　External links

- Official site

Chapter 2

Federal Emergency Management Agency & Related Articles

2.1 Community emergency response team

In the United States a **Community Emergency Response Team (CERT)** can refer to

- one of five federal programs promoted under the umbrella organization Citizen Corps, which is funded in part by the Stafford Act;

- an implementation of the federal CERT program, administered by a local sponsoring agency, which receives Stafford grant funding, and provides standardized training and an implementation framework to community members;

- an organization of volunteer emergency workers who have received specific training in basic disaster response skills, and who agree to supplement existing emergency responders in the event of a major disaster.

Sometimes programs and organizations take different names, such as **Neighborhood Emergency Response Team (NERT)**, or **Neighborhood Emergency Team (NET)**.

The concept of civilian auxiliaries is similar to civil defense, which has a longer history. The CERT concept differs because it includes nonmilitary emergencies, and is coordinated with all levels of emergency authorities, local to national, via an overarching incident command system.

2.1.1 CERT Organization

A local government agency, often a fire department or emergency management agency, agrees to sponsor CERT within its jurisdiction. The sponsoring agency liaises with, deploys and may train or supervise the training of CERT members. The sponsoring agency receives and disburses federal and state Citizen Corps grant funds allocated to its CERT program. Many sponsoring agencies employ a full-time community-service person as liaison to the CERT members. In some communities, the liaison is a volunteer and CERT member.

As people are trained and agree to join the community emergency response effort, a CERT is formed. Initial efforts may result in a team with only a few members from across the community. As the number of members grow, a single community-wide team may subdivide. Multiple CERTs are organized into a hierarchy of teams consistent with ICS principles. This follows the Incident Command System (ICS) principle of Span of control until the ideal distribution is achieved: one or more teams are formed at each neighborhood within a community.

A **Teen Community Emergency Response Team (TEEN CERT)**, or **Student Emergency Response Team (SERT)**, can be formed from any group of teens.[1] A Teen Cert can be formed as a school club, service organization, Venturing Crew, Explorer Post, or the training can be added to a school's graduation curriculum. Some CERTs form a club or service corporation, and recruit volunteers to perform training on behalf of the sponsoring agency. This reduces the financial and human resource burden on the sponsoring agency.

When not responding to disasters, CERTs may

- raise funds for emergency response equipment in their community;

- provide first-aid, crowd control or other services at community events;

- hold planning, training, or recruitment meetings; and

- conduct or participate in disaster response exercises.

A CERT volunteer practices using a fire extinguisher.

CERT volunteers try on their equipment.

Some sponsoring agencies use Citizen Corps grant funds to purchase response tools and equipment for their members and team(s) (subject to Stafford Act limitations). Most CERTs also acquire their own supplies, tools, and equipment. As community members, CERTs are aware of the specific needs of their community and equip the teams accordingly.

2.1.2   CERT Response

The basic idea is to use CERT to perform the large number of tasks needed in emergencies. This frees highly trained professional responders for more technical tasks. Much of CERT training concerns the Incident Command System and organization, so CERT members fit easily into larger command structures.

A team may self-activate (self-deploy) when their own neighborhood is affected by disaster. An effort is made to report their response status to the sponsoring agency. A self-activated team will size-up the loss in their neighborhood and begin performing the skills they have learned to minimize further loss of life, property, and environment. They will continue to respond safely until redirected or relieved by the sponsoring agency or professional responders on-scene.

Teams in neighborhoods not affected by disaster may be deployed or activated by the sponsoring agency. The sponsoring agency may communicate with neighborhood CERT leaders through an organic communication team. In some areas the communications may be by amateur radio, FRS, GMRS or MURS radio, dedicated telephone or fire-alarm networks. In other areas, relays of bicycle-equipped runners can effectively carry messages between the teams and the local emergency operations center.

The sponsoring agency may activate and dispatch teams in order to gather or respond to intelligence about an in-

cident. Teams may be dispatched to affected neighborhoods, or organized to support operations. CERT members may augment support staff at an Incident Command Post or Emergency Operations Center. Additional teams may also be created to guard a morgue, locate supplies and food, convey messages to and from other CERT teams and local authorities, and other duties on an as-needed basis as identified by the team leader.

In the short term, CERTs perform data gathering, especially to locate mass-casualties requiring professional response, or situations requiring professional rescues, simple fire-fighting tasks (for example, small fires, turning off gas), light search and rescue, damage evaluation of structures, triage and first aid. In the longer term, CERTs may assist in the evacuation of residents, or assist with setting up a neighborhood shelter.

While responding, CERT members are temporary volunteer government workers. In some areas, (such as California, Hawaii and Kansas) registered, activated CERT members are eligible for worker's compensation for on-the-job injuries during declared disasters.

2.1.3   CERT Member Roles

The Federal Emergency Management Agency (FEMA) recommends that the standard, ten-person team be comprised as follows:[2]

- **CERT Leader**. Generally, the first CERT team member arriving on the scene becomes team leader, and is the designated Incident Commander (IC) until the arrival of someone more competent. This person makes the IC initial assessment of the scene and determines the appropriate course of action for team members; assumes role of Safety Officer until assigned to another team member; assigns team member roles if not already assigned; designates triage area, treatment area, morgue, and vehicle traffic routes; coordinates and directs team operations; determines logistical needs (water, food, medical supplies, transportation, equipment, and so on.) and determines ways to meet those needs through team members or citizen volunteers on the scene; collects and writes reports on the operation and victims; and communicates and coordinates with the incident commander, local authorities, and other CERT team leaders. The team leader is identified by two pieces of crossed tape on the hard hat.

- **Safety Officer**. Checks team members prior to deployment to ensure they are safe and equipped for the operation; determines safe or unsafe working environments; ensures team accountability; supervises operations (when possible) where team members and victims are at direct physical risk, and alerts team members when unsafe conditions arise.

- **Fire Suppression Team (2 people)**. Work under the supervision of the Team Leader to suppress small fires in designated work areas or as needed; when not accomplishing their primary mission, assist the search and rescue team or triage team; assist in evacuation and transport as needed; assist in the triage or treatment area as needed, other duties as assigned; communicate with Team Leader.

- **Search and Rescue Team (2)**. Work under the supervision of the Team Leader, searching for and providing rescue of victims as is prudent under the conditions; when not accomplishing their primary mission, assist the Fire Suppression Team, assist in the triage or treatment area as needed; other duties as assigned; communicate with Team Leader.

- **Medical Triage Team (2)**. Work under the supervision of the Team Leader, providing START triage for victims found at the scene; marking victims with category of injury per the standard operating procedures; when not accomplishing their primary mission, assist the Fire Suppression Team if needed, assist the Search and Rescue Team if needed, assist in the Medical Triage Area if needed, assist in the Treatment Area if needed, other duties as assigned; communicate with Team Leader.

- **Medical Treatment Team (2)**. Work under the supervision of the Team Leader, providing medical treatment to victims within the scope of their training. This task is normally accomplished in the Treatment Area, however, it may take place in the affected area as well. When not accomplishing their primary mission, assist the Fire Suppression Team as needed,

An equipment trailer belonging to the Springfield, Illinois CERT program.

assist the Medical Triage Team as needed; other duties as assigned; communicate with the Team Leader.

Because every CERT member in a community receives the same core instruction, any team member has the training necessary to assume any of these roles. This is important during a disaster response because not all members of a regular team may be available to respond. Hasty teams may be formed by whichever members are responding at the time. Additionally, members may need to adjust team roles due to stress, fatigue, injury, or other circumstances.

2.1.4 CERT Training

While state and local jurisdictions will implement training in the manner that best suits the community, the Citizen Corps CERT program has an established curriculum. Jurisdictions may augment the training, but are strongly encouraged to deliver the entire core content. The Citizen Corps CERT core curriculum for the basic course is composed of the following nine units (time is instructional hours):[3]

- **Unit 1: Disaster Preparedness** (2.5 hrs). Topics include (in part) identifying local disaster threats, disaster impact, mitigation and preparedness concepts, and an overview of Citizen Corps and CERT. Hands on skills include team-building exercises, and shutting off utilities.

- **Unit 2: Fire Safety** (2.5 hrs). Students learn about fire chemistry, mitigation practices, hazardous materials identification, suppression options, and are introduced to the concept of size-up. Hands-on skills include using a fire extinguisher to suppress a live flame, and wearing basic protective gear.

- **Unit 3: Disaster Medical Operations part 1** (2.5 hrs). Students learn to identify and treat certain life-threatening conditions in a disaster setting, as well as START triage. Hands-on skills include performing head-tilt/chin-lift, practicing bleeding control techniques, and performing triage as an exercise.

- **Unit 4: Disaster Medical Operations part 2** (2.5 hrs). Topics cover mass casualty operations, public health, assessing patients, and treating injuries. Students practice patient assessment, and various treatment techniques.

- **Unit 5: Light Search and Rescue Operations** (2.5 hrs). Size-up is expanded as students learn about assessing structural damage, marking structures that have been searched, search techniques, as well as rescue techniques and cribbing. Hands-on activities include lifting and cribbing an object, and practicing rescue carries.

- **Unit 6: CERT Organization** (1.5 hrs). Students are introduced to several concepts from the Incident Command System, and local team organization and communication is explained. Hands-on skills include a table-top exercise focusing on incident command and control.

- **Unit 7: Disaster Psychology** (1 hr). Responder well-being and dealing with victim trauma are the topics of this unit. NOTE: Some programs such as Mid America TEEN CERT in Missouri teach **Special Needs Considerations** (2 hrs) that focuses on helping people with special needs or needing special/functional assistance.

- **Unit 8: Terrorism and CERT** (2.5 hrs). Students learn how terrorists may choose targets, what weapons they may use, and identifying when chemical, biological, radiological, nuclear, or explosive weapons may have been deployed. Students learn about CERT roles in preparing for and responding to terrorist attacks. A table-top exercise highlights topics covered.

- **Unit 9: Course Review and Disaster Simulation** (2.5 hrs). Students take a written exam, then participate in a real-time practical disaster simulation where the different skill areas are put to the test. A critique follows the exercise where students and instructors have an opportunity to learn from mistakes and highlight exemplary actions. Students may be given a certificate of completion at the conclusion of the course.

Citizen Corps CERT training emphasizes safely "doing the most good for the most people as quickly as possible" when responding to a disaster. For this reason, cardiopulmonary resuscitation (CPR) training is not included in the core curriculum, as it is time and responder intensive in a mass-casualty incident. However, many jurisdictions encourage or require CERT members to obtain CPR training. Many CERT programs provide or encourage members to take additional first aid training. Some CERT members may

also take training to become a certified first responder or emergency medical technician.

Arlington Community Emergency Response Team (CERT) provided back-up support to the county's 911 system.

Many CERT programs also provide training in amateur radio operation, shelter operations, flood response, community relations, mass care, the incident command system (ICS, and the National Incident Management System(NIMS).

Each unit of Citizen Corps CERT training is ideally delivered by professional responders or other experts in the field addressed by the unit. This is done to help build unity between CERT members and responders, keep the attention of students, and help the professional response organizations be comfortable with the training which CERT members receive.

Each course of instruction is ideally facilitated by one or more instructors certified in the CERT curriculum by the state or sponsoring agency. Facilitating instructors provide continuity between units, and help ensure that the CERT core curriculum is being delivered successfully. Facilitating instructors also perform set-up and tear-down of the classroom, provide instructional materials for the course, record student attendance and other tasks which assist the professional responder in delivering their unit as efficiently as possible.

Citizen Corps CERT training is provided free to interested members of the community, and is delivered in a group classroom setting. People may complete the training without obligation to join a CERT. Citizen Corps grant funds can be used to print and provide each student with a printed manual. Some sponsoring agencies use Citizen Corps grant funds to purchase disaster response tool kits. These kits are offered as an incentive to join a CERT, and must be returned to the sponsoring agency when members resign from CERT.

Some sponsoring agencies require a criminal background-check of all trainees before allowing them to participate on a CERT. For example, the city of Albuquerque, New Mexico require all volunteers to pass a background check,[4] while the city of Austin, Texas does not require a background check to take part in training classes but requires members to undergo a background check in order to receive a CERT badge and directly assist first reponders during an activation of the Emergency Operations Center in Austin.[5] However, most programs do not require a criminal background check in order to participate.

CERT volunteers during the classroom portion of their training.

The Citizen Corps CERT curriculum (including the Train-the-Trainer course) was updated during the last half of 2008 to reflect feedback from instructors across the nation. The update is in final review, and is scheduled for release during the first quarter of 2009.

2.1.5 See also

- Local Emergency Planning Committee

- Emergency management

- Incident command system

2.1.6 References

[1] "Teen CERT: Disaster Teen Training Guide" (PDF). *floridadisaster.org*. Volunteer USA Foundation. Retrieved 17 July 2015.

[2] "Standing Operating Procedures". The Woodlands CERT Committee. Retrieved 17 July 2015.

[3] Schmidt, James B. "A Community Emergency Response Team in the Marshfield Fire and Rescue Department" (PDF). U.S. Fire Administration. Retrieved 17 July 2015.

[4] "Community Emergency Response Team". *City of Albuquerque Official Website*. Retrieved 17 July 2015.

[5] "CERT Volunteers". City of Austin Office of Homeland Security and Emergency Management. Retrieved 17 July 2015.

2.1.7 External links

- U.S. Citizen Corps CERT

- Find a CERT Program

- Starting and Maintaining a CERT Organization: Resource Center.

- LAFD/ Los Angeles City CERT

- Spartanburg County SC CERT

- Mid America TEEN CERT

- Collegiate CERT from FEMA

- Southwest Johnson County CERT, Kansas

2.2 CRP-2B

CRP-2B (Crisis Relocation Program 2B) is a hypothetical scenario of nuclear war between the United States and the Soviet Union that was created in 1976 by the Federal Emergency Management Agency. It involved the detonation of 1444 weapons, with a yield of 6559 megatonnes, and projected a death toll of between 85 and 125 megadeaths.[1][2]

The program was referred to as a "study", but in fact it was the product of a computer simulation. It was also the source of an 80% survival rate figure that was quoted by many people in the years afterwards. The 80% survival rate was an initial assumption, built into the parameters of the computer simulation by its designers. But as the program came to be a "study", so the survival rate figure came to be the "finding" of the study.[1]

Charles F. Estes Jr., director of strategic police at the Office of the Undersecretary of Defence for Policy, stated:[1]

> It was assumed at the outset that 80 percent of the population of hypothesized target areas would in fact have been evacuated and would survive. [...] [T]his was an entering assumption rather than one of the study's analytically derived findings. [...] It is importance to note that the survivability assumptions [...] of the computer model were derived from opinions of interested civil defense program managers, academics, and contractor personnel These opinions were obtained through the use of accepted opinion survey techniques.
> — Charles F. Estes Jr., [1][3]

This 80% figure was quoted in the 1980s by U.S. congressmen and other officials.[1]

2.2.1 References

[1] Lee Ben Clarke (1999). *Mission Improbable*. University of Chicago Press. p. 36. ISBN 0-226-10941-0.

[2] Fredric Solomon and Robert Q. Marston (1986). *The Medical Implications of Nuclear War*. National Academies Press. p. 590. ISBN 0-309-03636-4.

[3] Jennifer Leaning and Langley Carleton Keyes (1983). *Counterfeit Ark: Crisis Relocation for Nuclear War*. Ballinger Pub. Co. pp. xix. ISBN 0-88410-940-2.

2.3 Disaster medical assistance team

For the baseball pitcher, see Daisuke Matsuzaka.

A **disaster medical assistance team** (**DMAT**) is a group

National Disaster Medical System logo.

of professional and para-professional medical personnel organized to provide rapid-response medical care or casualty decontamination during a terrorist attack, natural disaster, or other incident in the United States.

DMATs are part of the National Disaster Medical System and operate under the Department of Health and Human Services (DHHS).[1]

2.3.1 Organization

There are 80 NDMS Teams of which 55 are DMATs spread out across the country and are formed by local groups of health care providers and support personnel. Under the National Response Framework (NRF, DMATs are defined

according to their level of capability and experience. Once a level of training and proficiency has been shown, the higher level of priority is given to the team.

In addition to medical DMATs, there are other response teams that specialize in specific types of medical emergencies such as hazardous material handling and decontamination and LRATs, which are primarily Logistics response teams to support any of the response teams under the NDMS umbrella. Such other types of teams are the DMORTs (Disaster Mortuary Operations Response Teams, NVRTs (National Veterinary Medical Response Teams), IMSuRTs (International Medical/Surgical Response Teams), and the IRCTs (Incident Response Coordination Teams).

A DMAT deploys to disaster sites with the assurance by OPEO that supplies and equipment will arrive at or before the teams arrive at a disaster site, so that they can be self-sufficient for 72 hours while providing medical care at a fixed or temporary medical care site. Responsibilities may include triaging patients, providing high-quality medical care in adverse and austere environments, and preparing patients for evacuation. Other situations may involve providing primary medical care or augmenting overloaded local health care facilities and staffs. DMATs have been used to implement mass inoculations and other immediate needs to large populations. Under rare circumstances, disaster victims may be evacuated to a different locale to receive medical care. DMATs may be activated to support patient reception and distribution of patients to hospitals.

2.3.2 Team composition and equipment

DMATs are composed of physicians, nurse practitioners, physician assistants, nurses, pharmacists and pharmacy technicians, respiratory therapists, paramedics, Emergency Medical Technicians, and a variety of other health and logistical personnel. DMATs typically have 120–150 members, from which the Team Leader chooses up to 50 members to deploy on missions requiring a full team. Smaller strike teams or other modular units can also be activated and deployed when less than full-scale deployments are needed.

DMAT members are termed "intermittent" federal employees and once activated by federal order, their status changes to that of an active federal employee and follow the GS pay scale. Federally activated DMAT members are protected from tort liability while in operation and are also protected by the provisions of the Uniformed Services Employment and Reemployment Rights Act (USERRA) which affords the same protections extended to National Guard and Active Duty Military when they deploy in that their full-time jobs are not placed in jeopardy. This protection was not until 2003 after an act of congress.

DMATs formerly traveled equipped with medical equipment and supplies, large tents, generators, and other support equipment (cache) necessary to establish a Base of Operations, designed to be self-sufficient for up to 72 hours, in a disaster area and treat up to 250 patients per day. However, during the period 2009–2011, ASPR changed the operational tactics and has removed team caches to a small number of federal warehouses, to save money, and thus teams no longer have the opportunity to practice and train with their own caches as was done previously. The capability is similar to an urgent care-level health care facility. In 2005, FEMA increased the response capabilities of DMATs by issuing trucks to teams that have obtained a certain standard of training and capabilities, but these, too, have been reclaimed by ASPR and are only available during actual deployments, to deliver the caches from the federal warehouses.

2.3.3 Incidents

DMATs have been called to respond to a variety incidents, many of which garnered significant media attention. Teams responded to the World Trade Center site in New York City and the Pentagon, following the 9/11 terrorist attacks.

DMATs are a critical element of the federal response to natural disasters including Hurricane Katrina, During Katrina DMAT teams treated and helped evacuate patients in and around New Orleans, including those at the Louisiana Superdome and Louis Armstrong New Orleans International Airport.

Between January 17 and February 22, 2010, twelve DMATs participated in the international response to the 2010 Haiti earthquake and cared for more than 31,300 patients, including 167 surgeries and the delivery of 45 infants.

More recently, DMATs have aided in the response to Hurricane Sandy, which was particularly devastating to areas of New York and New Jersey.

2.3.4 References

[1] "DMAT". National Disaster Medical System. Retrieved September 11, 2012.

2.3.5 External links

- "National Disaster Medical System". Retrieved September 11, 2012.

- "HHS.gov". U.S. Department of Health & Human Services. Retrieved September 11, 2012.

- "USERRA". US Office of Special Counsel. Retrieved September 11, 2012.

- http://www.oh-1dmat.org

2.4 Disaster Mortuary Operational Response Team

A **Disaster Mortuary Operational Response Team** or **DMORT** is a team of experts in the fields of victim identification and mortuary services. DMORTs are activated in response to large scale disasters in the United States to assist in the identification of deceased individuals and storage of the bodies pending the bodies being claimed.

2.4.1 Organization

For organizational purposes, the country is divided into ten regions, each with a Regional Coordinator.[1] For the duration of their service, DMORT members work under the local authorities of the disaster site and their professional licenses are recognized by all states.

DMORT Teams:

- REGION I (ME, NH, VT, MA, CT, RI)

- REGION II (NY, NJ, PR, VI)

- REGION III (PA, MD, DC, DE, VA, WV)

- REGION IV (AL, KY, TN, NC, SC, GA, MS, FL)

- REGION V (MN, WI, IL, IN, MI, OH)

- REGION VI (NM, TX, OK, AR, LA)

- REGION VII (NE, IA, KS, MO)

- REGION VIII (MT, ND, SD, WY, UT, CO)

- REGION IX (AZ, NV, CA, HI)

- REGION X (WA, AK, OR, ID)

DMORTs are organized under the Department of Health and Human Services National Disaster Medical System. The DMORTs are composed of civilian funeral directors, medical examiners, coroners, pathologists, forensic anthropologists, fingerprint specialists, forensic odontologists, dental assistants, and radiographers. They are supported by medical records technicians and transcribers, mental health specialists, computer professionals, administrative support staff, and security and investigative personnel. When a

DMORT is activated, the personnel on the team are treated and paid as a temporary Federal employee.[1]

The Department of Health and Human Services maintains three Disaster Portable Morgue Units (DPMU) which are staged at HHS Logistics Centers, one each in Frederick, Maryland, Fort Worth, Texas, and San Jose, California. Each DPMU is a cache of equipment and supplies for a complete morgue with designated workstations for each process the DMORT team is required to complete.[1]

2.4.2 History

In the 1980s, the National Funeral Directors Association (NFDA) formed a committee to address the need for a way of dealing with mass casualty situations. The group had the goal of formulating a plan for funeral directors to deal with the situation. As the committee worked on the plan, it was revealed that such a situation would call for multiple forensic specialties. The committee created the first portable morgue unit in the country.[2]

The committee's work came to the attention of the Federal Government following the complaints of families whose family members had been lost in airline incidents. The families felt that the remains hadn't received adequate treatment. The United States Congress passed the Family Assistance Act in 1996.[2] The National Transportation Safety Board (NTSB) was assigned the role managing the Federal response to aviation disaster victims and their families. The division responsible for this response was the Office of Family Affairs, later renamed the Office of Transportation Disaster Assistance. The NTSB made use of DMORTs to handle large scale transportation disasters.[3]

Following the creation of the Department of Homeland Security in 2002, the DMORTs were moved into the Emergency Preparedness and Response directorate as part of the National Disaster Medical System.[4] In 2007 the National Disaster Medical System was removed from DHS and returned to the Department of Health and Human Services under the control of the Assistant Secretary for Preparedness and Response. April 8, 2011 http://www.phe.gov/Preparedness/responders/ndms/teams/Pages/dmort.aspx

2.4.3 Identification of remains

It is a two-part process that utilizes a sophisticated computer program for matching physical characteristics. The families of the deceased provide as much information about them as possible: dental records, x-rays, photographs or descriptions of tattoos, clothing and jewelry; blood type information and objects that may contain the deceased's DNA, such as hair or a toothbrush. The information gathered,

called antemortem, or "before death" information, is entered into a computer program called VIP (Victim Identification Profile), which is capable of assimilating 800 different item categories, including graphics, photographs and x-rays. As forensic scientists (pathologists, anthropologists, odontologists) examine the recovered remains, they enter their findings - called postmortem data—into VIP (Victim Identification Profile). Depending on the availability of data, the WIN-VIP system enables scientists to match the remains to their identity.

2.4.4 Incidents

For the World Trade Center disaster, U.S. Secretary of Health and Human Services Tommy G. Thompson activated the National Disaster Medical System. It was the first time this federally coordinated response system had been activated on a full nationwide basis.

In 2006, DMORT operated the Find Family National Call Center in Baton Rouge, Louisiana. This is the center of all operations concerning the location and reuniting of families scattered by Hurricane Katrina and Hurricane Rita. Out of nearly 13,000 people reported missing after the impacts of hurricanes Katrina, Rita, Stan, and Wilma, nearly 7,000 were found alive and reunited with their families.

2.4.5 References

[1] "DMORT". National Disaster Medical System. Retrieved September 6, 2006.

[2] "DMORT History Page". DMORT.org. February 28, 2004. Retrieved September 7, 2006.

[3] "The NTSB and DMORT". DMORT.org. April 23, 2005. Retrieved September 7, 2006.

[4] "Who Became Part of the Department?". *History of the Department of Homeland Security*. Department of Homeland Security. Archived from the original on September 5, 2006. Retrieved September 7, 2006.

2.4.6 External links

- Disaster Mortuary Operational Response Teams (DMORTs)
- DMORT Mass Fatality Assistance

2.5 Emergency management

This article is about community plans in the event of a disaster. For business preparedness, see Crisis management.

Disaster management (or **emergency management**) is the creation of plans through which communities reduce vulnerability to hazards and cope with disasters.[1][2] Disaster management does not avert or eliminate the threats, instead it focuses on creating plans to decrease the impact of disasters. Failure to create a plan could lead to damage to assets, human mortality, and lost revenue. Currently in the United States 60% businesses do not have emergency management plans. Events covered by disaster management include acts of terrorism, industrial sabotage, fire, natural disasters (such as earthquakes, hurricanes, etc.), public disorder, industrial accidents, and communication failures.

2.5.1 Emergency planning ideals

If possible, emergency planning should aim to prevent emergencies from occurring, and failing that, should develop a good action plan to mitigate the results and effects of any emergencies. As time goes on, and more data becomes available, usually through the study of emergencies as they occur, a plan should evolve. The development of emergency plans is a cyclical process, common to many risk management disciplines, such as Business Continuity and Security Risk Management, as set out below:

- Recognition or identification of risks

- Ranking or evaluation of risks

  - Responding to significant risks

  - Tolerate

  - Treat

  - Transfer

  - Terminate

- Resourcing controls

- Reaction Planning

- Reporting & monitoring risk performance

- Reviewing the Risk Management framework

There are a number of guidelines and publications regarding Emergency Planning, published by various professional organisations such as ASIS, FEMA and the Emergency Planning College. There are very few Emergency Management specific standards, and emergency management as a discipline tends to fall under business resilience standards.

In order to avoid, or reduce significant losses to a business, emergency managers should work to identify and anticipate potential risks, hopefully to reduce their probability of occurring. In the event that an emergency does occur, managers should have a plan prepared to mitigate the effects of that emergency, as well as to ensure Business Continuity of critical operations post-incident. It is essential for an organisation to include procedures for determining whether an emergency situation has occurred and at what point an emergency management plan should be activated.

2.5.2 Implementation ideals

An emergency plan must be regularly maintained, in a structured and methodical manner, to ensure it is up-to-date in the event of an emergency. Emergency managers generally follow a common process to anticipate, assess, prevent, prepare, respond and recover from an incident.

Pre-incident training and testing

A team of emergency responders performs a training scenario involving anthrax.

Emergency management plans and procedures should include the identification of appropriately trained staff members responsible for decision-making when an emergency occurs. Training plans should include internal people, contractors and civil protection partners, and should state the nature and frequency of training and testing.

Testing of a plan's effectiveness should occur regularly. In instances where several business or organisations occupy the same space, joint emergency plans, formally agreed to by all parties, should be put into place.

Communicating and incident assessment

Communication is one of the key issues during any emergency, pre-planning of communications is critical. Miscommunication can easily result in emergency events escalating unnecessarily.

Once an emergency has been identified a comprehensive assessment evaluating the level of impact and its financial implications should be undertaken. Following assessment, the appropriate plan or response to be activated will depend on a specific pre-set criteria within the emergency plan. The steps necessary should be prioritized to ensure critical functions are operational as soon as possible.

2.5.3   Phases and personal activities

Emergency management consists of five phases: prevention, mitigation, preparedness, response and recovery.

Prevention

Prevention was recently added to the phases of emergency management. It focuses on preventing the human hazard, primarily from potential natural disasters or terrorist attacks. Preventive measures are taken on both the domestic and international levels, designed to provide permanent protection from disasters. Not all disasters, particularly natural disasters, can be prevented, but the risk of loss of life and injury can be mitigated with good evacuation plans, environmental planning and design standards. In January 2005, 168 Governments adopted a 10-year global plan for natural disaster risk reduction called the Hyogo Framework.

Mitigation

Personal mitigation is a key to national preparedness. Individuals and families train to avoid unnecessary risks. This includes an assessment of possible risks to personal/family health and to personal property, and steps taken to minimize the effects of a disaster, or take procure insurance to protect them against effects of a disaster.

Preventive or mitigation measures take different forms for different types of disasters. In earthquake prone areas, these preventive measures might include structural changes such as the installation of an earthquake valve to instantly shut off the natural gas supply, seismic retrofits of property, and the securing of items inside a building. The latter may include the mounting of furniture, refrigerators, water heaters and breakables to the walls, and the addition of cabinet latches. In flood prone areas, houses can be built on poles/stilts. In areas prone to prolonged electricity black-outs installation

of a generator. The construction of storm cellars and fallout shelters are further examples of personal mitigative actions.

On a national level, governments might implement large scale mitigation measures. After the monsoon floods of 2010, the Punjab government subsequently constructed 22 'disaster-resilient' model villages, comprising 1885 single-storey homes, together with schools and health centres.[3]

Preparedness

An airport emergency preparedness exercise

Preparedness focuses on preparing equipment and procedures for use when a disaster occurs. This equipment and these procedures can be used to reduce vulnerability to disaster, to mitigate the impacts of a disaster or to respond more efficiently in an emergency. The Federal Emergency Management Agency (FEMA) has set out a basic four-stage vision of preparedness flowing from mitigation to preparedness to response to recovery and back to mitigation in a circular planning process.[4] This circular, overlapping model has been modified by other agencies, taught in emergency class and discussed in academic papers.[5] FEMA also operates a Building Science Branch that develops and produces multi-hazard mitigation guidance that focuses on creating disaster-resilient communities to reduce loss of life and property.[6] FEMA advises citizens to prepare their homes with some emergency essentials in the case that the food distribution lines are interrupted. FEMA has subsequently prepared for this contingency by purchasing hundreds of thousands of freeze dried food emergency meals ready to eat (MRE's) to dispense to the communities where emergency shelter and evacuations are implemented.

Emergency preparedness can be difficult to measure.[7] CDC focuses on evaluating the effectiveness of its public health efforts through a variety of measurement and assessment programs.[8]

Local Emergency Planning Committees

Local Emergency Planning Committees (LEPCs) are required by the United States Environmental Protection Agency under the Emergency Planning and Community Right-to-Know Act to develop an emergency response plan, review the plan at least annually, and provide information about chemicals in the community to local citizens.[9] This emergency preparedness effort focuses on hazards presented by use and storage of extremely hazardous, hazardous and toxic chemicals.[10] Particular requirements of LEPCs include

- Identification of facilities and transportation routes of extremely hazardous substances

- Description of emergency response procedures, on and off site

- Designation of a community coordinator and facility emergency coordinator(s) to implement the plan

- Outline of emergency notification procedures

- Description of how to determine the probable affected area and population by releases

- Description of local emergency equipment and facilities and the persons responsible for them

- Outline of evacuation plans

- A training program for emergency responders (including schedules)

- Methods and schedules for exercising emergency response plans

According to the EPA, "Many LEPCs have expanded their activities beyond the requirements of EPCRA, encouraging accident prevention and risk reduction, and addressing homeland security in their communities" and the Agency offers advice on how to evaluate the effectiveness of these committees.[11]

Preparedness measures

Preparedness measures can take many forms ranging from focusing on individual people, locations or incidents to broader, government-based "all hazard" planning.[12] There are a number of preparedness stages between "all hazard' and individual planning, generally involving some combination of both mitigation and response planning. Business continuity planning encourages businesses to have a Disaster Recovery Plan. Community- and faith-based organizations mitigation efforts promote field response teams and inter-agency planning.[13]

Classroom response kit

School-based response teams cover everything from live shooters to gas leaks and nearby bank robberies.[14] Educational institutions plan for cyberattacks and windstorms.[15] Industry specific guidance exists for horse farms,[16] boat owners[17] and more.

Family preparedness for disaster is fairly unusual. A 2013 survey found that only 19% of American families felt that they were "very prepared" for a disaster.[18] Still, there are many resources available for family disaster planning. The Department of Homeland Security's Ready.gov page includes a Family Emergency Plan Checklist,[19] has a whole webpage devoted to readiness for kids, complete with cartoon-style superheroes,[20] and ran a Thunderclap Campaign in 2014.[21] The Center for Disease Control has a Zombie Apocalypse website.[22]

Kitchen fire extinguisher

Disasters take a variety of forms to include earthquakes, tsunamis or regular structure fires. That a disaster or emergency is not large scale in terms of population or acreage impacted or duration does not make it any less of a disaster for the people or area impacted and much can be learned about preparedness from so-called small disasters.[23] The Red Cross states that it responds to nearly 70,000 disasters a year, the most common of which is a single family fire.[24]

Items on shelves in basement

Preparedness starts with an individual's everyday life and involves items and training that would be useful in an emergency. What is useful in an emergency is often also useful in everyday life as well.[25] From personal preparedness, preparedness continues on a continuum through family preparedness, community preparedness and then business, non-profit and governmental preparedness. Some organizations blend these various levels. For example, the International Red Cross and Red Crescent Movement has a webpage on disaster training[26] as well as offering training on basic preparedness such as Cardiopulmonary resuscitation and First Aid. Other non-profits such as Team Rubicon bring specific groups of people into disaster preparedness and response operations.[27] FEMA breaks down preparedness into a pyramid, with citizens on the foundational bottom, on top of which rests local government, state government and federal government in that order.[28]

The basic theme behind preparedness is to be ready for an emergency and there are a number of different variations of being ready based on an assessment of what sort of threats exist. Nonetheless, there is basic guidance for preparedness that is common despite an area's specific dangers. FEMA

Non-perishable food in cabinet

recommends that everyone have a three-day survival kit for their household.[29] Because individual household sizes and specific needs might vary, FEMA's recommendations are not item specific, but the list includes:

- Three-day supply of non-perishable food.

- Three-day supply of water – one gallon of water per person, per day.

- Portable, battery-powered radio or television and extra batteries.

- Flashlight and extra batteries.

- First aid kit and manual.

- Sanitation and hygiene items (moist towelettes and toilet paper).

- Matches and waterproof container.

- Whistle.

- Extra clothing.

- Kitchen accessories and cooking utensils, including a can opener.

- Photocopies of credit and identification cards.

- Cash and coins.

- Special needs items, such as prescription medications, eyeglasses, contact lens

- solutions, and hearing aid batteries.

- Items for infants, such as formula, diapers, bottles, and pacifiers.

- Other items to meet unique family needs.

Along similar lines, but not exactly the same, CDC has its own list for a proper disaster supply kit.[30]

- Water—one gallon per person, per day

- Food—nonperishable, easy-to-prepare items

- Flashlight

- Battery powered or hand crank radio (NOAA Weather Radio, if possible)

- Extra batteries

- First aid kit

- Medications (7-day supply), other medical supplies, and medical paperwork (e.g., medication list and pertinent medical information)

- Multipurpose tool (e.g., Swiss army knife)

- Sanitation and personal hygiene items

- Copies of personal documents (e.g., proof of address, deed/lease to home, passports, birth certificates, and insurance policies)

- Cell phone with chargers

- Family and emergency contact information

- Extra cash

- Emergency blanket

- Map(s) of the area

- Extra set of car keys and house keys

- Manual can opener

Children are a special population when considering Emergency preparedness and many resources are directly focused on supporting them. SAMHSA has list of tips for talking to children during infectious disease outbreaks, to include being a good listener, encouraging children to ask questions and modeling self-care by setting routines, eating healthy meals, getting enough sleep and taking deep breaths

to handle stress.[31] FEMA has similar advice, noting that "Disasters can leave children feeling frightened, confused, and insecure" whether a child has experienced it first hand, had it happen to a friend or simply saw it on television.[32] In the same publication, FEMA further notes, "Preparing for disaster helps everyone in the family accept the fact that disasters do happen, and provides an opportunity to identify and collect the resources needed to meet basic needs after disaster. Preparation helps; when people feel prepared, they cope better and so do children."

To help people assess what threats might be in order to augment their emergency supplies or improve their disaster response skills, FEMA has published a booklet called the "Threat and Hazard Identification and Risk Assessment Guide."[33] (THIRA) This guide, which outlines the THIRA process, emphasizes "whole community involvement," not just governmental agencies, in preparedness efforts. In this guide, FEMA breaks down hazards into three categories: Natural, technological and human caused and notes that each hazard should be assessed for both its likelihood and its significance. According to FEMA, "Communities should consider only those threats and hazards that could plausibly occur" and "Communities should consider only those threats and hazards that would have a significant effect on them." To develop threat and hazard context descriptions, communities should take into account the time, place, and conditions in which threats or hazards might occur.

Not all preparedness efforts and discussions involve the government or established NGOs like the Red Cross. Emergency preparation discussions are active on the internet, with many blogs and websites dedicated to discussing various aspects of preparedness. On-line sales of items such as survival food, medical supplies and heirloom seeds allow people to stock basements with cases of food and drinks with 25 year shelf lives, sophisticated medical kits and seeds that are guaranteed to sprout even after years of storage.[34]

One group of people who put a lot of effort in disaster preparations is called Doomsday Preppers. This subset of preparedness-minded people often share a belief that the FEMA or Red Cross emergency preparation suggestions and training are not extensive enough. Sometimes called survivalists, Doomsday Preppers are often preparing for The End Of The World As We Know It, abbreviated as TEOTWAWKI. With a motto some have that "The Future Belongs to those who Prepare," this Preparedness subset has its own set of Murphy's Rules,[35] including "Rule Number 1: Food, you still don't have enough" and "Rule Number 26: People who thought the Government would save them, found out that it didn't."

Not all emergency preparation efforts revolve around food, guns and shelters, though these items help address the needs in the bottom two sections of Maslow's hierarchy of needs.

The American Preppers Network[36] has an extensive list of items that might be useful in less apparent ways than a first aid kid or help add 'fun' to challenging times. These items include:

- Books and magazines

- Arts and crafts

- Children's entertainment

- Crayons and coloring books

- Notebooks and writing supplies

- Nuts, bolts, screws, nails, etc.

- Religious material

- Sporting equipment, card games and board games

Emergency preparedness goes beyond immediate family members. For many people, pets are an integral part of their families and emergency preparation advice includes them as well. It is not unknown for pet owners to die while trying to rescue their pets from a fire or from drowning.[37] CDC's Disaster Supply Checklist for Pets includes:[30]

- Food and water for at least 3 days for each pet; bowls, and a manual can opener.

- Depending on the pet you may need a litter box, paper towels, plastic trash bags, grooming items, and/or household bleach.

- Medications and medical records stored in a waterproof container.

- First aid kit with a pet first aid book.

- Sturdy leash, harness, and carrier to transport pet safely. A carrier should be large enough for the animal to stand comfortably, turn around, and lie down. Your pet may have to stay in the carrier for several hours.

- Pet toys and the pet's bed, if you can easily take it, to reduce stress.

- Current photos and descriptions of your pets to help others identify them in case you and your pets become separated, and to prove that they are yours.

- Information on feeding schedules, medical conditions, behavior problems, and the name and telephone number of your veterinarian in case you have to board your pets or place them in foster care.

Emergency preparedness also includes more than physical items and skill-specific training. Psychological preparedness is also a type of emergency preparedness and specific mental health preparedness resources are offered for mental health professionals by organizations such as the Red Cross.[24] These mental health preparedness resources are designed to support both community members affected by a disaster and the disaster workers serving them. CDC has a website devoted to coping with a disaster or traumatic event.[38] After such an event, the CDC, through the Substance Abuse and Mental Health Services Administration (SAMHSA), suggests that people seek psychological help when they exhibit symptoms such as excessive worry, crying frequently, an increase in irritability, anger, and frequent arguing, wanting to be alone most of the time, feeling anxious or fearful, overwhelmed by sadness, confused, having trouble thinking clearly and concentrating, and difficulty making decisions, increased alcohol and/or substance use, increased physical (aches, pains) complaints such as headaches and trouble with "nerves."

Sometimes emergency supplies are kept in what is called a Bug-out bag. While FEMA does not actually use the term "Bug out bag," calling it instead some variation of a "Go Kit," the idea of having emergency items in a quickly accessible place is common to both FEMA and CDC, though on-line discussions of what items a "bug out bag" should include sometimes cover items such as firearms and great knives that are not specifically suggested by FEMA or CDC.[39] The theory behind a "bug out bag" is that emergency preparations should include the possibility of Emergency evacuation. Whether fleeing a burning building or hastily packing a car to escape an impending hurricane, flood or dangerous chemical release, rapid departure from a home or workplace environment is always a possibility and FEMA suggests having a Family Emergency Plan for such occasions.[40] Because family members may not be together when disaster strikes, this plan should include reliable contact information for friends or relatives who live outside of what would be the disaster area for household members to notify they are safe or otherwise communicate with each other. Along with the contact information, FEMA suggests having well-understood local gathering points if a house must be evacuated quickly to avoid the dangers of re-reentering a burning home.[41] Family and emergency contact information should be printed on cards and put in each family member's backpack or wallet. If family members spend a significant amount of time in a specific location, such as at work or school, FEMA suggests learning the emergency preparation plans for those places.[40] FEMA has a specific form, in English and in Spanish, to help people put together these emergency plans, though it lacks lines for email contact information.[40]

Like children, people with disabilities and other special

needs have special emergency preparation needs. While "disability" has a specific meaning for specific organizations such as collecting Social Security benefits,[42] for the purposes of emergency preparedness, the Red Cross uses the term in a broader sense to include people with physical, medical, sensor or cognitive disabilities or the elderly and other special needs populations.[43] Depending on the particular disability, specific emergency preparations might be required. FEMA's suggestions for people with disabilities includes having copies of prescriptions, charging devices for medical devices such as motorized wheel chairs and a week's supply of medication readily available LINK or in a "go stay kit."[44] In some instances, lack of competency in English may lead to special preparation requirements and communication efforts for both individuals and responders.[45]

FEMA notes that long term power outages can cause damage beyond the original disaster that can be mitigated with emergency generators or other power sources to provide an Emergency power system.[46] The United States Department of Energy states that 'homeowners, business owners, and local leaders may have to take an active role in dealing with energy disruptions on their own."[47] This active role may include installing or other procuring generators that are either portable or permanently mounted and run on fuels such as propane or natural gas[48] or gasoline.[49] Concerns about carbon monoxide poisoning, electrocution, flooding, fuel storage and fire lead even small property owners to consider professional installation and maintenance.[46] Major institutions like hospitals, military bases and educational institutions often have or are considering extensive backup power systems.[50] Instead of, or in addition to, fuel-based power systems, solar, wind and other alternative power sources may be used.[51] Standalone batteries, large or small, are also used to provide backup charging for electrical systems and devices ranging from emergency lights to computers to cell phones.[52]

Emergency preparedness does not stop at home or at school.[53] The United States Department of Health and Human Services addresses specific emergency preparedness issues hospitals may have to respond to, including maintaining a safe temperature, providing adequate electricity for life support systems and even carrying out evacuations under extreme circumstances.[54] FEMA encourages all businesses to have businesses to have an emergency response plan[55] and the Small Business Administration specifically advises small business owners to also focus emergency preparedness and provides a variety of different worksheets and resources.[56]

FEMA cautions that emergencies happen while people are travelling as well[57] and provides guidance around emergency preparedness for a range travelers to include commuters,[58] *Commuter Emergency Plan* and holiday travelers.[59] In particular, Ready.gov has a number of emergency preparations specifically designed for people with cars.[60] These preparations include having a full gas tank, maintaining adequate windshield wiper fluid and other basic car maintenance tips. Items specific to an emergency include:

- Jumper cables: might want to include flares or reflective triangle

- Flashlights, to include extra batteries (batteries have less power in colder weather)

- First Aid Kit, to include any necessary medications, baby formula and diapers if caring for small children

- Non-perishable food such as canned food (be alert to liquids freezing in colder weather), and protein rich foods like nuts and energy bars

- Manual can opener

- At least 1 gallon of water per person a day for at least 3 days (be alert to hazards of frozen water and resultant container rupture)

- Basic toolkit: pliers, wrench, screwdriver

- Pet supplies: food and water

- Radio: battery or hand cranked

- For snowy areas: cat litter or sand for better tire traction; shovel; ice scraper; warm clothes, gloves, hat, sturdy boots, jacket and an extra change of clothes

- Blankets or sleeping bags

- Charged Cell Phone: and car charger

In addition to emergency supplies and training for various situations, FEMA offers advice on how to mitigate disasters. The Agency gives instructions on how to retrofit a home to minimize hazards from a Flood, to include installing a Backflow prevention device, anchoring fuel tanks and relocating electrical panels.[61]

Given the explosive danger posed by natural gas leaks, Ready.gov states unequivocally that "It is vital that all household members know how to shut off natural gas" and that property owners must ensure they have any special tools needed for their particular gas hookups. Ready.gov also notes that "It is wise to teach all responsible household members where and how to shut off the electricity," cautioning that individual circuits should be shut off before the main circuit. Ready.gov further states that "It is vital that all household members learn how to shut off the water at the main house valve" and cautions that the possibility that rusty valves might require replacement.[62]

Marked gas shuttoff

Evacuation sign

Response

The response phase of an emergency may commence with Search and Rescue but in all cases the focus will quickly turn to fulfilling the basic humanitarian needs of the affected population. This assistance may be provided by national or international agencies and organizations. Effective coordination of disaster assistance is often crucial, particularly when many organizations respond and local emergency management agency (LEMA) capacity has been exceeded by the demand or diminished by the disaster itself. The National Response Framework is a United States government publication that explains responsibilities and expectations of government officials at the local, state, federal, and tribal levels. It provides guidance on Emergency Support Functions that may be integrated in whole or parts to aid in the response and recovery process.

On a personal level the response can take the shape either of a *shelter in place* or an *evacuation*.

In a shelter-in-place scenario, a family would be prepared to fend for themselves in their home for many days without any form of outside support. In an *evacuation*, a family leaves the area by automobile or other mode of transportation, taking with them the maximum amount of supplies they can carry, possibly including a tent for shelter. If mechani-

cal transportation is not available, evacuation on foot would ideally include carrying at least three days of supplies and rain-tight bedding, a tarpaulin and a bedroll of blankets.

Donations are often sought during this period, especially for large disasters that overwhelm local capacity. Due to efficiencies of scale, money is often the most cost-effective donation if fraud is avoided. Money is also the most flexible, and if goods are sourced locally then transportation is minimized and the local economy is boosted. Some donors prefer to send gifts in kind, however these items can end up creating issues, rather than helping. One innovation by Occupy Sandy volunteers is to use a donation registry, where families and businesses impacted by the disaster can make specific requests, which remote donors can purchase directly via a web site.

Medical considerations will vary greatly based on the type of disaster and secondary effects. Survivors may sustain a multitude of injuries to include lacerations, burns, near drowning, or crush syndrome.

Recovery

The recovery phase starts after the immediate threat to human life has subsided. The immediate goal of the recovery phase is to bring the affected area back to normalcy as quickly as possible. During reconstruction it is recommended to consider the location or construction material of the property.

The most extreme home confinement scenarios include war, famine and severe epidemics and may last a year or more. Then recovery will take place inside the home. Planners for these events usually buy bulk foods and appropriate storage and preparation equipment, and eat the food as part of normal life. A simple balanced diet can be constructed from vitamin pills, whole-meal wheat, beans, dried milk, corn, and cooking oil.[63] One should add vegetables, fruits, spices and meats, both prepared and fresh-gardened, when possible.

2.5.4 As a profession

Professional emergency managers can focus on government and community preparedness, or private business preparedness. Training is provided by local, state, federal and private organizations and ranges from public information and media relations to high-level incident command and tactical skills.

In the past, the field of emergency management has been populated mostly by people with a military or first responder background. Currently, the field has become more diverse, with many managers coming from a variety of backgrounds other than the military or first responder fields. Educational opportunities are increasing for those seeking undergraduate and graduate degrees in emergency management or a related field. There are over 180 schools in the US with emergency management-related programs, but only one doctoral program specifically in emergency management.[64]

Professional certifications such as Certified Emergency Manager (CEM)[65] and Certified Business Continuity Professional (CBCP) are becoming more common as professional standards are raised throughout the field, particularly in the United States. There are also professional organizations for emergency managers, such as the National Emergency Management Association and the International Association of Emergency Managers.

Principles

In 2007, Dr. Wayne Blanchard of FEMA's Emergency Management Higher Education Project, at the direction of Dr. Cortez Lawrence, Superintendent of FEMA's Emergency Management Institute, convened a working group of emergency management practitioners and academics to consider principles of emergency management. This was the first time the principles of the discipline were to be cod-

ified. The group agreed on eight principles that will be used to guide the development of a doctrine of emergency management. Below is a summary:

1. Comprehensive – consider and take into account all hazards, all phases, all stakeholders and all impacts relevant to disasters.

2. Progressive – anticipate future disasters and take preventive and preparatory measures to build disaster-resistant and disaster-resilient communities.

3. Risk-driven – use sound risk management principles (hazard identification, risk analysis, and impact analysis) in assigning priorities and resources.

4. Integrated – ensure unity of effort among all levels of government and all elements of a community.

5. Collaborative – create and sustain broad and sincere relationships among individuals and organizations to encourage trust, advocate a team atmosphere, build consensus, and facilitate communication.

6. Coordinated – synchronize the activities of all relevant stakeholders to achieve a common purpose.

7. Flexible – use creative and innovative approaches in solving disaster challenges.

8. Professional – value a science and knowledge-based approach; based on education, training, experience, ethical practice, public stewardship and continuous improvement.

A fuller description of these principles can be found at[66]

Tools

In recent years the continuity feature of emergency management has resulted in a new concept, Emergency Management Information Systems (EMIS). For continuity and inter-operability between emergency management stakeholders, EMIS supports an infrastructure that integrates emergency plans at all levels of government and non-government involvement for all four phases of emergencies. In the healthcare field, hospitals utilize the Hospital Incident Command System (HICS), which provides structure and organization in a clearly defined chain of command.

Disaster response technologies

Smart Emergency Response System (SERS)[67] prototype was built in the SmartAmerica Challenge 2013-2014, a

United States government initiative. SERS has been created by a team of nine organizations led by MathWorks. The project was featured at the White House in June 2014 and described by Todd Park (U.S. Chief Technology Officer) as an exemplary achievement.

The Smart America initiative challenges the participants to build cyber-physical systems as a glimpse of the future to save lives, create jobs, foster businesses, and improve the economy. SERS primarily saves lives. The system provides the survivors and the emergency personnel with information to locate and assist each other during a disaster. SERS allows to submit help requests to a MATLAB-based mission center connecting first responders, apps, search-and-rescue dogs, a 6-feet-tall humanoid, robots, drones, and autonomous aircraft and ground vehicles. The command and control center optimizes the available resources to serve every incoming requests and generates an action plan for the mission. The Wi-Fi network is created on the fly by the drones equipped with antennas. In addition, the autonomous rotorcrafts, planes, and ground vehicles are simulated with Simulink and visualized in a 3D environment (Google Earth) to unlock the ability to observe the operations on a mass scale.[68]

2.5.5 Within other professions

A disaster plan book at Rockefeller University in a biochemistry research laboratory.

Practitioners in emergency management come from an increasing variety of backgrounds. Professionals from memory institutions (e.g., museums, historical societies, etc.) are dedicated to preserving cultural heritage—objects and records. This has been an increasingly major component within this field as a result of the heightened awareness following the September 11 attacks in 2001, the hurricanes in 2005, and the collapse of the Cologne Archives.

To increase the potential successful recovery of valuable

records, a well-established and thoroughly tested plan must be developed. This plan should emphasize simplicity in order to aid in response and recovery: employees should perform similar tasks in the response and recovery phase that they perform under normal conditions. It should also include mitigation strategies such as the installation of sprinklers within the institution.[69] Professional associations hold regular workshops to keep individuals up to date with tools and resources in order to minimize risk and maximize recovery.

Other tools

In 2008, the U.S. Agency for International Development created a web-based tool for estimating populations impacted by disasters. Called Population Explorer[70] the tool uses land scan population data, developed by Oak Ridge National Laboratory, to distribute population at a resolution 1 km^2 for all countries in the world. Used by USAID's FEWS NET Project to estimate populations vulnerable and or imd by food insecurity, Population Explorer is gaining wide use in a range of emergency analysis and response actions, including estimating populations impacted by floods in Central America and the Pacific Ocean Tsunami event in 2009.

In 2007, a checklist for veterinarians was published in the Journal of the American Veterinary Medical Association, it had two sets of questions for a professional to ask themselves before assisting with an emergency:

Absolute requirements for participation:

- Have I chosen to participate?
- Have I taken ICS training?
- Have I taken other required background courses?
- Have I made arrangements with my practice to deploy?
- Have I made arrangements with my family?

Incident Participation:

- Have I been invited to participate
- Are my skill sets a match for the mission?
- Can I access just-in-time training to refresh skills or acquire needed new skills?
- Is this a self-support mission?
- Do I have supplies needed for three to five days of self-support?

While written for veterinarians, this checklist is applicable for any professional to consider before assisting with an emergency.[71]

2.5.6 International organizations

The International Emergency Management Society

The International Emergency Management Society (TIEMS), is an international non-profit NGO, registered in Belgium. TIEMS is a Global Forum for Education, Training, Certification and Policy in Emergency and Disaster Management. TIEMS' goal is to develop and bring modern emergency management tools, and techniques into practice, through the exchange of information, methodology innovations and new technologies.

TIEMS provides a platform for stakeholders to meet, network and learn about new technical and operational methodologies. TIEMS focuses on cultural differences to be understood and included in the society's events, education and research programs. This is achieved by establishing local chapters worldwide. Today, TIEMS has chapters in Benelux, Romania, Finland, Italy, Middle East and North Africa (MENA), Iraq, India, Korea, Japan and China.

International Association of Emergency Managers

The International Association of Emergency Managers (IAEM) is a non-profit educational organization aimed at promoting the goals of saving lives and property protection during emergencies. The mission of IAEM is to serve its members by providing information, networking and professional opportunities, and to advance the emergency management profession.

It has seven councils around the world: Asia,[72] Canada,[73] Europa,[74] International,[75] Oceania,[76] Student[77] and USA.[78]

The Air Force Emergency Management Association, affiliated by membership with the IAEM, provides emergency management information and networking for U.S. Air Force Emergency Management personnel.

International Recovery Platform

The International Recovery Platform (IRP) was conceived at the World Conference on Disaster Reduction (WCDR) in Kobe, Hyogo, Japan in January 2005, as part of the *Hyogo Framework for Action (HFA) 2005–2015*. The HFA is a global plan for disaster risk reduction adopted by 168 governments.

The key role of IRP is to identify gaps in post disaster recovery and to serve as a catalyst for the development of tools and resources for recovery efforts.[79]

The International Red Cross and Red Crescent Movement

The International Federation of Red Cross and Red Crescent Societies (IFRC) works closely with National Red Cross and Red Crescent societies in responding to emergencies, many times playing a pivotal role. In addition, the IFRC may deploy assessment teams, e.g. Field Assessment and Coordination Teams (FACT),[80] to the affected country if requested by the national society. After assessing the needs, Emergency Response Units (ERUs)[81] may be deployed to the affected country or region. They are specialized in the response component of the emergency management framework.

Baptist Global Response

Baptist Global Response (BGR) is a disaster relief and community development organization. BGR and its partners respond globally to people with critical needs world-wide, whether those needs arise from chronic conditions or acute crises such as natural disasters. While BGR is not an official entity of the Southern Baptist Convention, it is rooted in Southern Baptist life and is the international partnership of Southern Baptist Disaster Relief teams, which operate primarily in the US and Canada.[82]

United Nations

The United Nations system rests with the Resident Coordinator within the affected country. However, in practice, the UN response will be coordinated by the UN Office for the Coordination of Humanitarian Affairs (UN-OCHA), by deploying a UN Disaster Assessment and Coordination (UNDAC) team, in response to a request by the affected country's government.

World Bank

Since 1980, the World Bank has approved more than 500 projects related to disaster management, dealing with both disaster mitigation as well as reconstruction projects, amounting to more than US$40 billion. These projects have taken place all over the world, in countries such as Argentina, Bangladesh, Colombia, Haiti, India, Mexico, Turkey and Vietnam.[83][83]

Prevention and mitigation projects include forest fire prevention measures, such as early warning measures and education campaigns; early-warning systems for hurricanes; flood prevention mechanisms (e.g. shore protection, terracing, etc.); and earthquake-prone construction.[83] In a joint venture with Columbia University under the umbrella of the ProVention Consortium the World Bank has established a Global Risk Analysis of Natural Disaster Hotspots.[84]

In June 2006, the World Bank, in response to the HFA, established the Global Facility for Disaster Reduction and Recovery (GFDRR), a partnership with other aid donors to reduce disaster losses. GFDRR helps developing countries fund development projects and programs that enhance local capacities for disaster prevention and emergency preparedness.[85]

European Union

In 2001 the EU adopted Community Mechanism for Civil Protection, to facilitate co-operation in the event of major emergencies requiring urgent response actions. This also applies to situations where there may be an imminent threat as well.[86]

The heart of the Mechanism is the Monitoring and Information Center (MIC), part of the European Commission's Directorate-General for Humanitarian Aid & Civil Protection. Accessible 24 hours a day, it gives countries access to a one-stop-shop of civil protections available amongst all the participating states. Any country inside or outside the Union affected by a major disaster can make an appeal for assistance through the MIC. It acts as a communication hub, and provides useful and updated information on the actual status of an ongoing emergency.[87]

2.5.7   National organizations

Australia

Main article: Emergency Management in Australia

Natural disasters are part of life in Australia. Heatwaves have killed more Australians than any other type of natural disaster in the 20th century. Australia's emergency management processes embrace the concept of the prepared community. The principal government agency in achieving this is Emergency Management Australia.

Canada

Public Safety Canada is Canada's national emergency management agency. Each province is required to have both leg-islation for dealing with emergencies, and provincial emergency management agencies, typically called "Emergency Measures Organizations" (EMO). Public Safety Canada coordinates and supports the efforts of federal organizations as well as other levels of government, first responders, community groups, the private sector, and other nations. The Public Safety and Emergency Preparedness Act defines the powers, duties and functions of PS are outlined. Other acts are specific to individual fields such as corrections, law enforcement, and national security.

Germany

In Germany the Federal Government controls the German *Katastrophenschutz* (disaster relief), the Technisches Hilfswerk (*Federal Agency for Technical Relief*, THW), and the *Zivilschutz* (civil protection) programs coordinated by the *Federal Office of Civil Protection and Disaster Assistance*. Local fire department units, the German Armed Forces (Bundeswehr), the German Federal Police and the 16 state police forces (Länderpolizei) are also deployed during disaster relief operations.

There are several private organizations in Germany that also deal with emergency relief. Among these are the German Red Cross, Johanniter-Unfall-Hilfe (the German equivalent of the St. John Ambulance), the Malteser-Hilfsdienst, and the Arbeiter-Samariter-Bund. As of 2006, there is a program of study at the University of Bonn leading to the degree "Master in Disaster Prevention and Risk Governance"[88]

India

A protective wall built on the shore of the coastal town of Kalpakkam, in aftermath of the 2004 Indian Ocean earthquake.

The National Disaster Management Authority is the primary government agency responsible for planning and capacity-building for disaster relief. Its emphasis is primarily on strategic risk management and mitigation, as well as developing policies and planning.[89] The National Institute of Disaster Management is a policy think-tank and training institution for developing guidelines and training programs for mitigating disasters and managing crisis response.

The National Disaster Response Force is the government agency primarily responsible for emergency management during natural and man-made disasters, with specialized skills in search, rescue and rehabilitation.[90] The Ministry of Science and Technology also contains an agency that brings the expertise of earth scientists and meteorologists to emergency management. The Indian Armed Forces also plays an important role in the rescue/recovery operations after disasters.

Aniruddha's Academy of Disaster Management (ACDM) is a non-profit organization in Mumbai, India with 'disaster management' as its principal objective.

New Zealand

In New Zealand, depending on the scope of the emergency/disaster, responsibility may be handled at either the local or national level. Within each region, local governments are organized into 16 Civil Defence Emergency Management Groups (CMGs). If local arrangements are overwhelmed, pre-existing mutual-support arrangements are activated. Central government has the authority to coordinate the response through the National Crisis Management Centre (NCMC), operated by the Ministry of Civil Defence & Emergency Management (MCDEM). These structures are defined by regulation,[91] and explained in *The Guide to the National Civil Defence Emergency Management Plan 2006*, roughly equivalent to the U.S. Federal Emergency Management Agency's National Response Framework.

New Zealand uses unique terminology for emergency management. Emergency management is rarely used, many government publications retaining the use of the term civil defence.[92][93][94] For example, the Minister of Civil Defence is responsible for the MCDEM. Civil Defence Emergency Management is a term in its own right, defined by statute.[95] And disaster rarely appears in official publications, emergency and incident being the preferred terms,[96] with the term event also being used. For example, publications refer to the Canterbury Snow Event 2002[97]

- **4Rs** is the emergency management cycle used in New Zealand, its four phases are known as:[98]

  - Reduction = Mitigation
  - Readiness = Preparedness
  - Response

- Recovery

Pakistan

Disaster management in Pakistan revolves around flood disasters focusing on rescue and relief. There is a dearth of knowledge and information about hazard identification, risk assessment and management, and disaster preparedness. Disaster management, development planning and environmental management institutions operate in isolation with no integrated planning, there being no central authority for integrated disaster management. State-level measures are heavily tilted towards structural aspects.

Russia

In Russia, the Ministry of Emergency Situations (EMERCOM) is engaged in fire fighting, civil defense, and search and rescue after both natural and human-made disasters.

Somalia

In Somalia, the Federal Government announced in May 2013 that the Cabinet had approved draft legislation on a new Somali Disaster Management Agency (SDMA), which had originally been proposed by the Ministry of Interior. According to the Prime Minister's Media Office, the SDMA will lead and coordinate the government's response to various natural disasters. It is part of a broader effort by the federal authorities to re-establish national institutions. The Federal Parliament is now expected to deliberate on the proposed bill for endorsement after any amendments.[99]

The Netherlands

In the Netherlands the Ministry of Security and Justice is responsible for emergency preparedness and emergency management on a national level and operates a national crisis centre (NCC). The country is divided into 25 safety regions (veiligheidsregio). In a safety region, there are four components: the regional fire department, the regional department for medical care(ambulances and psycho-sociological care etc.), the regional dispatch and a section for risk- and crisis management. The regional dispatch operates for police, fire department and the regional medical care. The dispatch has all these three services combined into one dispatch for the best multi-coordinated response to an incident or an emergency. An also facilitates in information management, emergency communication and care of citizens. These services are the main structure for a response to an emergency. It can happen that, for a specific emergency, the co-operation with an other service is needed, for instance the Ministry of Defence, water board(s) or Rijkswaterstaat. The veiligheidsregio can integrate these other services into their structure by adding them to specific conferences on operational or administrative level.

All regions operate according to the Coordinated Regional Incident Management system.

United Kingdom

Following the 2000 fuel protests and severe flooding that same year, as well as the foot-and-mouth crisis in 2001, the United Kingdom passed the Civil Contingencies Act 2004 (CCA). The CCA defined some organisations as Category 1 and 2 Responders, setting responsibilities regarding emergency preparedness and response. It is managed by the Civil Contingencies Secretariat through Regional Resilience Forums and local authorities.

Disaster Management training is generally conducted at the local level, and consolidated through professional courses that can be taken at the Emergency Planning College. Diplomas, undergraduate and postgraduate qualifications can be gained at universities throughout the country. The Institute of Emergency Management is a charity, established in 1996, providing consulting services for the government, media and commercial sectors. There are a number of professional societies for Emergency Planners including the Emergency Planning Society[100] and the Institute of Civil Protection and Emergency Management.[101]

One of the largest emergency exercises in the UK was carried out on 20 May 2007 near Belfast, Northern Ireland: a simulated plane crash-landing at Belfast International Airport. Staff from five hospitals and three airports participated in the drill, and almost 150 international observers assessed its effectiveness.[102]

United States

Disaster management in the United States has utilized the functional All-Hazards approach for over 20 years, in which managers develop processes (such as communication & warning or sheltering) rather than developing single-hazard or threat focused plans (e.g., a tornado plan). Processes are then mapped to specific hazards or threats, with the manager looking for gaps, overlaps, and conflicts between processes.

Given these notions, emergency managers must identify, contemplate, and assess possible man-made threats and natural threats that may affect their respective locales.[103] Because of geographical differences throughout the nation, a variety of different threats affect communities among the states. Thus, although similarities may exist, no two

emergency plans will be completely identical. Additionally, each locale has different resources and capacities (e.g., budgets, personnel, equipment, etc.) for dealing with emergencies.[104] Each individual community must craft its own unique emergency plan that addresses potential threats that are specific to the locality.[105]

This creates a plan more resilient to unique events because all common processes are defined, and encourages planning done by the stakeholders who are closer to the individual processes, such as a traffic management plan written by public works director. This type of planning can lead to conflict with non-emergency management regulatory bodies, which require development of hazard/threat specific plans, such as development of specific H1N1 flu plans and terrorism-specific plans.

In the United States, all disasters are initially local, with local authorities, with usually a police, fire, or EMS agency, taking charge. Many local municipalities may also have a separate dedicated office of emergency management (OEM), along with personnel and equipment. If the event becomes overwhelming to local government, state emergency management (the primary government structure of the United States) becomes the controlling emergency management agency. Federal Emergency Management Agency (FEMA), part of the Department of Homeland Security (DHS), is lead federal agency for emergency management. The United States and its territories are broken down into ten regions for FEMA's emergency management purposes. FEMA supports, but does not override, state authority.

The Citizen Corps is an organization of volunteer service programs, administered locally and coordinated nationally by DHS, which seek to mitigate disasters and prepare the population for emergency response through public education, training, and outreach. Most disaster response is carried out by volunteer organizations. In the US, the Red Cross is chartered by Congress to coordinate disaster response services, including typically being the lead agency handling shelter and feeding of evacuees. Religious organizations, with their ability to provide volunteers quickly, are usually integral during the response process. The largest being the Salvation Army,[106] with a primary focus on chaplaincy and rebuilding, and Southern Baptists who focus on food preparation and distribution,[107] as well as cleaning up after floods and fires, chaplaincy, mobile shower units, chainsaw crews and more. With over 65,000 trained volunteers Southern Baptist Disaster Relief is one of the largest disaster relief organizations in the US.[108] Similar services are also provided by Methodist Relief Services, the Lutherans, and Samaritan's Purse. Unaffiliated volunteers show up at most large disasters. To prevent abuse by criminals and for the safety of the volunteers, procedures have been implemented within most response agencies to manage and effectively use these 'SUVs' (Spontaneous Unaffiliated Volunteers).[109]

The US Congress established the Center for Excellence in Disaster Management and Humanitarian Assistance (COE) as the principal agency to promote disaster preparedness in the Asia-Pacific region.

The National Tribal Emergency Management Council (NEMC) is a non-profit educational organization developed for Tribal organizations to share information and best practices, as well as discussing issues regarding public health and safety, emergency management and homeland security, affecting those under Indian sovereignty. NTMC is organized into Regions, based on the FEMA 10 region system. NTMC was founded by the Northwest Tribal Emergency Management Council (NWTEMC), a consortium of 29 Tribal Nations and Villages in Washington, Idaho, Oregon and Alaska.

If a disaster or emergency is declared to be terror related or an "Incident of National Significance", the Secretary of Homeland Security will initiate the National Response Framework (NRF). The NRF allows the integration of federal resources with local, county, state, or tribal entities, with management of those resources to be handled at the lowest possible level, utilizing the National Incident Management System (NIMS).

FEMA's Emergency Management Institute See also: Emergency Management Institute
The Emergency Management Institute (EMI) serves as the

Emergency Management Institute's Main Campus in Emmitsburg, Maryland

national focal point for the development and delivery of emergency management training to enhance the capabilities of state, territorial, local, and tribal government officials; volunteer organizations; FEMA's disaster workforce; other Federal agencies; and the public and private sectors to minimize the impact of disasters and emergencies on the American public. EMI curricula are structured to meet the needs of this diverse audience with an emphasis on separate

organizations working together in all-hazards emergencies to save lives and protect property. Particular emphasis is placed on governing doctrine such as the National Response Framework (NRF), National Incident Management System (NIMS), and the National Preparedness Guidelines.[110] EMI is fully accredited by the International Association for Continuing Education and Training (IACET) and the American Council on Education (ACE).[111]

Approximately 5,500 participants attend resident courses each year while 100,000 individuals participate in nonresident programs sponsored by EMI and conducted by state emergency management agencies under cooperative agreements with FEMA. Another 150,000 individuals participate in EMI-supported exercises, and approximately 1,000 individuals participate in the Chemical Stockpile Emergency Preparedness Program (CSEPP).[112]

The *independent study* program at EMI consists of free courses offered to United States citizens in Comprehensive Emergency Management techniques.[113] Course IS-1 is entitled "Emergency Manager: An Orientation to the Position" and provides background information on FEMA and the role of emergency managers in agency and volunteer organization coordination. The EMI Independent Study (IS) Program, a Web-based distance learning program open to the public, delivered extensive online training with approximately 200 courses and trained more than 2.8 million individuals. The EMI IS Web site receives 2.5 to 3 million visitors a day.[114]

2.5.8 See also

- Disaster medicine
- Water security and emergency preparedness
- Rohn Emergency Scale
- Public health emergency (United States)
- Emergency Communication System
- ISITEP
- Mass fatality incident
- Liquidator (Chernobyl)

NGOs:

- Catholic Relief Services[115]
- Consortium of British Humanitarian Agencies
- Disaster Accountability Project (DAP)
- GlobalMedic

- Humanitarian International Services Group (HISG)

- International Disaster Emergency Service (IDES)

- Médecins Sans Frontières

- NetHope

2.5.9   References

[1] "Maine Emergency Management Agency" (2007). "What is Emergency Management?". Retrieved 2014-02-22.

[2] Drabek, Thomas (1991). *Emergency Management: Principles and Practice for Local Government.* Washington, D.C.: International City Management Association. pp. xvii.

[3] *NEWS: Pakistan's Punjab builds model villages to withstand disasters*, Climate & Development Knowledge Network, 17 December 2013.

[4] "Animals in Disasters". Training.fema.gov. Retrieved 2015-03-06.

[5] Baird, Malcolm E. (2010). *"The "Phases" of Emergency Management"* (PDF). Vanderbilt Center for Transportation Research. Retrieved 2015-03-08.

[6] "Building Science". *fema.gov.* Retrieved 8 March 2015.

[7] "Rand Homeland Security" (PDF). Rand.org. Retrieved 2015-03-08.

[8] "Public Health Emergency Preparedness Cooperative Agreement" (PDF). Cdc.gov. Retrieved 2015-03-08.

[9] "US Environmental Protection Agency | US EPA". .epa.gov. 2015-01-28. Retrieved 2015-03-08.

[10] "The Emergency Planning and Community Right-to-Know Act" (PDF). Epa.gov. Retrieved 2015-03-08.

[11] "Masuring Progress in Chemical Safety : A Guide for Local Emergency Planning Committees and Similar Groups" (PDF). Epa.gov. Retrieved 2015-03-08.

[12] http://www.fema.gov/pdf/plan/slg101.pdf *Guide for All-Hazard Emergency Operations* Planning September 1996

[13] "Community-Based Pre-Disaster Mitigation" (PDF). Fema.gov. Retrieved 2015-03-08.

[14] "School-Based Emergency preparedness : A National analysis and recommended Protocol" (PDF). Archive.ahrq.gov. Retrieved 2015-03-08.

[15] AHRQ Publication No. 09-0013 January 2009 *UMass System Office Hazard Mitigation Plan Draft (December 2013)*

[16] "Horse Farm Disaster Preparedness". TheHorse.com. 2014-11-26. Retrieved 2015-03-08.

[17]

[18] "Your family's emergency kit is probably a disaster". Cnn.com. Retrieved 2015-03-08.

[19] "Family Emergency Plan" (PDF). Ready.gov. Retrieved 2015-03-08.

[20] "Home". Ready.gov. Retrieved 2015-03-08.

[21] "Thunderclap: Resolve to be Ready in 2014!". Thunderclap.it. 2013-12-18. Retrieved 2015-03-08.

[22] "Preparedness 101: Zombie Apocalypse | Public Health Matters Blog | Blogs | CDC". Blogs.cdc.gov. 2011-05-16. Retrieved 2015-03-08.

[23]

[24] "Disaster Mental Health : Introduction". a1881.g.akamai.net. Retrieved 2015-03-08.

[25] "Everyday Objects for Travel Disaster Survival". Travel Insurance Review. Retrieved 2015-03-08.

[26] "Disaster Training". *redcross.org.* Retrieved 8 March 2015.

[27] "Program Ops". *Disaster Response Veterans Service Organization - Team Rubicon.* Retrieved 8 March 2015.

[28] "Why Prepare" (PDF). Fema.gov. Retrieved 2015-03-08.

[29] "Basic Preparedness" (PDF). Fema.gov. Retrieved 2015-03-08.

[30] "CDC Emergency preparedness and You | Gather Emergency Supplies | Disaster Supplies Kit". Emergency.cdc.gov. Retrieved 2015-03-08.

[31] "Talking With Children" (PDF). Store.samhsa.gov. Retrieved 2015-03-08.

[32] "Coping with Disaster". FEMA.gov. 2015-01-31. Retrieved 2015-03-08.

[33] "Threat and Hazrd Identification and Risk Assessment Guide" (PDF). Fema.gov. August 2013. Retrieved 2015-03-08.

[34] "My Patriot Supply : Homepage". Mypatriotsupply.com. Retrieved 2015-03-06.

[35] "Murphy's Laws of TEOTWAWKI". Survivalcache.com. 2010-11-16. Retrieved 2015-03-08.

[36] "American Preppers Network - National family survival and preparedness organization". *American Preppers Network.* Retrieved 8 March 2015.

[37] "Man dies trying to rescue pets from fire". Pottsmerc.com. Retrieved 2015-03-08.

[38] "Coping with Traumatic Events, SAMHSA.gov". Media.samhsa.gov. 2011-08-25. Retrieved 2015-03-08.

[39] "Bug Out Bag Checklist". Theprepperproject.com. 2013-04-18. Retrieved 2015-03-08.

[40] "Make A Plan". Ready.gov. 2014-01-29. Retrieved 2015-03-08.

[41] "Escape From Fire!" (PDF). Usfa.fema.gov. Retrieved 2015-03-08.

[42] "Who Will Qualify For Disability? - What Qualifying Is Based On". Ssdrc.com. Retrieved 2015-03-08.

[43] "Preparing for Disaster for People with Disabilities and other Special Needs" (PDF). Redcross.org. Retrieved 2015-03-08.

[44] (archived link, February 11, 2015)

[45] "Ready New York : Preparing for Emergencies in new York City" (PDF). Nyc.com. Retrieved 2015-03-08.

[46] "Install a Generator for Emergency Power" (PDF). Fema.gov. Retrieved 2015-03-08.

[47] "Community Guidelines for Energy Emergencies | Department of Energy". Energy.gov. Retrieved 2015-03-08.

[48] Generac Power Systems, Inc. "Generac Home Backup Power - Home & Portable Generator - Generac Power Systems". *generac.com*. Retrieved 8 March 2015.

[49] "Duromax RV Grade 4,400-Watt 7.0 HP Gasoline Powered Portable Generator with Wheel Kit-XP4400 - The Home Depot". *The Home Depot*. 15 October 2014. Retrieved 8 March 2015.

[50] "Microgrid Effects and Opportunities for Utilities" (PDF). Burnsmcd.com. Retrieved 2015-03-08.

[51] "Alternative Energy Sources For Homes During Emergencies". House Hold Power Generator. 2013-08-18. Retrieved 2015-03-08.

[52] "Goal Zero Yeti 400". *Crutchfield*. Retrieved 8 March 2015.

[53] "Plan for Locations". Ready.gov. 2014-01-30. Retrieved 2015-03-08.

[54] "Hospital Evacuation Decision Guide: Chapter 2. Pre-Disaster Self-Assessment". Archive.ahrq.gov. 2011-06-30. Retrieved 2015-03-08.

[55] "Emergency Response Plan". Ready.gov. 2012-12-19. Retrieved 2015-03-08.

[56] https://www.sba.gov/content/disaster-preparedness *Emergency Preparedness*

[57] "FEMA.gov Communities - National Preparedness Community Main Group - Travel Preparedness Tips: How to Be Ready When On The Go". Community.fema.gov. 2014-07-31. Retrieved 2015-03-08.

[58] "Commuter Emergency Plan" (PDF). Fema.gov. Retrieved 2015-03-06.

[59] "Planes, Trains, and Automobiles - Holiday Travel Safety Tips". FEMA.gov. 2012-06-18. Retrieved 2015-03-08.

[60] "Car Safety". Ready.gov. Retrieved 2015-03-08.

[61] "Homeowner's Guide to Retrofitting" (PDF). Fema.gov. Retrieved 2015-03-08.

[62] "Utility Shut-off & Safety". Ready.gov. Retrieved 2015-03-08.

[63] "Federal Emergency Management Agency". FEMA.gov. Retrieved 2013-08-11.

[64] Jaffin, Bob (September 17, 2008). "Emergency Management Training: How to Find the Right Program". Emergency Management Magazine. Retrieved 2008-11-15.

[65] "Certification-General CEM certification Info". Iaem.com. Retrieved 2012-03-07.

[66] "Principles of Emergency Management Supplement" (PDF). 2007-09-11. Retrieved 2015-03-06.

[67] "Smart Emergency Response System (SERS)". Smart America. Retrieved 2015-03-08.

[68] "The Smart Emergency Response System Using MATLAB and Simulink". YouTube. 2014-07-18. Retrieved 2015-03-08.

[69] Buchanan, Sally. "Emergency preparedness." from Paul Banks and Roberta Pilette. *Preservation Issues and Planning*. Chicago: American Library Association, 2000. 159–165. ISBN 978-0-8389-0776-4

[70] "FEWS Network — USAID". Population Explorer. Retrieved 2012-03-07.

[71] The Veterinary profession's duty of care in response to disasters and food animal emergencies. *Journal of the American Veterinary Medical Association*, Vol 231, No. 2, July 15, 2007

[72] "IAEMAsia". Iaem.com. Retrieved 2012-03-07.

[73] "Region 13". Iaem.com. Retrieved 2012-03-07.

[74] "IAEM Europa". Iaem.com. Retrieved 2012-03-07.

[75]

[76] "IAEM Oceania". Iaem.com. Retrieved 2012-03-07.

[77] "Welcome to IAEM.COM- International Association of Emergency Managers". Iaem.com. Retrieved 2012-03-07.

[78] "Iaem-Usa". Iaem.com. Retrieved 2012-03-07.

[79] "Welcome to the International Recovery Platform — International Recovery Platform". Recoveryplatform.org. Retrieved 2015-01-15.

[80] "Field Assessment Coordination Teams (FACT)". IFRC. 2011-10-15. Retrieved 2012-03-07.

[81] "Emergency Response Units (ERUs)". IFRC. 2011-10-15. Retrieved 2012-03-07.

[82] https://gobgr.org/about/about

[83] "Disaster Risk Management - Projects". Web.worldbank.org. 2004-04-28. Retrieved 2015-03-08.

[84] "Core Data Sets". Ldeo.columbia.edu. Retrieved 2015-03-08.

[85]

[86] Boin, A.; Rhinard, M. (2008). "Managing Transboundary Crises: What Role for the European Union?". *International Studies Review* **10**: 1. doi:10.1111/j.1468-2486.2008.00745.x.

[87] "Civil Protection – The Community mechanism for civil protection". Ec.europa.eu. Retrieved 2010-07-29.

[88] Marc Jansen (2010-06-29). "Startseite des Studiengangs Katastrophenvorsorge und -management". Kavoma.de. Retrieved 2010-07-29.

[89] "Functions and Responsibilities". National Disaster Management Authority. Retrieved 2014-10-28.

[90] "About Us". National Disaster Response Force. Retrieved 2015-01-15.

[91] National Civil Defence Emergency Plan Order 2005. Legislation.govt.nz (2008-10-01). Retrieved on 2011-07-28.

[92] "Civil Defence". beehive.govt.nz. Retrieved 2015-03-08.

[93]

[94]

[95] "Civil Defence Emergency Management Act 2002 No 33 (as at 01 January 2014), Public Act 4 Interpretation – New Zealand Legislation". Legislation.govt.nz. Retrieved 2015-03-08.

[96] For example, disaster is not used in the Civil Defence Emergency Management Act 2002, the enabling legislation for New Zealand's emergency management

[97] Retrieved 3 August according to rahul jain the fludes and natural uncertainties are included in mgt it is known as disaster mgt2008. (PDF) . Retrieved on 2011-07-28.

[98] *National Civil Defence Emergency Management Strategy 2007*, page 5. Department of Internal Affairs, Wellington, New Zealand 2008. Digital edition. Retrieved 3 August 2008. ISBN 0-478-29453-0.

[99] Prime Minister's Media Office (30 May 2013). "SOMALIA: Prime Minister calls for parliament to enact legislation as Cabinet moves to establish disaster management agency". *Raxanreeb*. Retrieved 5 June 2013.

[100] "Welcome - The Emergency Planning Society". Theeps.org. Retrieved 2015-03-08.

[101] "Institute of Civil Protection & Emergency Management | Welcome". ICPEM. 2014-04-03. Retrieved 2015-03-08.

[102] "UK | Northern Ireland | Mock plane crash tests NI crews". BBC News. 2007-05-20. Retrieved 2015-03-08.

[103] McElreath, David; Doss, Daniel; Jensen, Carl; Wigginton, Michael; Nations, Robert; Van Slyke, Jeffrey; Nations, Julie (2014). *Foundations of Emergency Management* (1st ed.). Dubuque, IA: Kendall-Hunt Publishing Company. p. 25. ISBN 978-1465234889.

[104] Doss, Daniel; Glover, William; Goz, Rebecca; Wigginton, Michael (2015). *The Foundations of Communication in Criminal Justice Systems*. Boca Raton, Florida: CRC Press. p. 301. ISBN 978-1482236576.

[105] Doss, Daniel; Glover, William; Goza, Rebecca; Wigginton, Michael (2015). *The Foundations of Communication in Criminal Justice Systems* (1 ed.). Boca Raton, Florida: CRC Press. p. 301. ISBN 978-1482236576.

[106] "The Salvation Army Emergency Disaster Services". Disaster.salvationarmyusa.org. Retrieved 2015-03-08.

[107] "2012 Activity Report". Namb.net. Retrieved 2015-03-08.

[108] http://www.baptistrelief.org/Our_Work/

[109] "Citizen Corps | Ready.gov" (PDF). Citizencorps.gov. 2013-07-23. Retrieved 2013-08-11.

[110] http://training.fema.gov/

[111] EMI Accredited

[112] EMI Overview

[113] FEMA EMI Independent Study home

[114] IS Course Offerings

[115] "Emergency Response | Catholic Relief Services". Crs.org. Retrieved 2015-03-08.

2.5.10   Further reading

- International Journal of Emergency Management, ISSN 1741-5071 (electronic) ISSN 1471-4825 (paper), Inderscience Publishers

- Journal of Homeland Security and Emergency Management ISSN 1547-7355, Bepress

- Australian Journal of Emergency Management (electronic) ISSN 1324-1540 (paper), Emergency Management Australia

- Karanasios, S. (2011). New and Emergent ICTs and Climate Change in Developing Countries. In R. Heeks & A. Ospina (Eds.). Manchester: Centre for Development Informatics, University of Manchester

- The ALADDIN Project, a consortium of universities developing automated disaster management tools

- Emergency Management Australia (2003) *Community Developments in Recovering from Disaster*, Commonwealth of Australia, Canberra

- Plan and Prep: Surviving the Zombie Apocalypse, (paperback), CreateSpace, Introductory concepts to planning and preparing for emergencies and disasters of any kind.

2.5.11 External links

- *Emergency Management Training*

- *Emergency Management Australia*

- *Disaster Plan Workbook*

- *The Disaster Mitigation Planning Assistance Website.*

- Public Health Management after Natural Disasters: Preparation, Response & Recovery – video, presentations, and summary of event held at the Woodrow Wilson International Center for Scholars, June 2008

- Emergency Response Resources The National Institute for Occupational Safety and Health

- FAO in emergencies

- Resilient Livelihoods: Disaster Risk Reduction for Food and Nutrition Security - 2013 edition published by FAO

2.6 Emergency Management Institute

EMI's Main Campus within the National Emergency Training Center in Emmitsburg.

The **Emergency Management Institute** (EMI) of the United States Federal Emergency Management Agency serves as the national focal point for the development and delivery of emergency management training to enhance the capabilities of state, territorial, local, and tribal government officials; volunteer organizations; FEMA's disaster workforce; other Federal agencies; and the public and private sectors to minimize the impact of disasters and emergencies on the American public. EMI curricula are structured to meet the needs of this diverse audience with an emphasis on separate organizations working together in all-hazards emergencies to save lives and protect property. Particular emphasis is placed on governing doctrine such as the National Response Framework (NRF), National Incident Management System (NIMS), and the National Preparedness Guidelines.[1] EMI is fully accredited by the International Association for Continuing Education and Training (IACET) and the American Council on Education (ACE).[2] The instruction is based upon the principles of Emergency Management and instructional systems design, which create a framework within whole communities to reduce vulnerability to hazards and to cope with disasters. EMI develops courses and implements training delivery systems to include residential onsite training; off-site delivery in partnership with Emergency Management training systems, colleges, and universities; and technology-based mediums to conduct individual training courses for Emergency Management and Response personnel across the United States.

EMI provides the following services:

- Emergency Management certification and leadership training

- Mission assistance to organizations and teams

- Online knowledge-sharing resources

- Continuous learning assets

2.6.1 Main campus facilities

EMI's main campus is located within the National Emergency Training Center (NETC) in Emmitsburg, Maryland. NETC is located 12 miles south of Gettysburg, Pennsylvania, 75 miles north of Washington, DC, and 50 miles northwest of Baltimore, Maryland. The 107-acre campus is shared by the United States Fire Administration (USFA), the National Fire Academy (NFA), the Field Personnel Operations Division, and the Satellite Procurement Office. All these components are part of FEMA, one of the four directorates in DHS. The campus has fully equipped air-conditioned classrooms, lodging for students, a Learning Resource Center, and dining and recreational facilities.

There also are several specialized facilities, such as the Simulation and Exercise Lab, a television studio Preparedness Network (PREPnet), and two computer laboratories that are integral to the instruction of many courses.[3]

The UK counterpart of the Emergency Management Institute is the Emergency Planning College in Easingwold.

2.6.2   History

Seal of the FCDA

President Jimmy Carter

EMI first started as the Civil Defense Staff College (CDSC) in Olney, Maryland, on April 1, 1951, and taught civil defense courses in program administration and finance, radiation monitoring and control, and heavy rescue. Due to concerns during the Cold War the CDSC's parent organization, the Federal Civil Defense Administration (FCDA), under Administrator Val Peterson, saw the Presidential Order to move the FCDA and the CDSC to Battle Creek, Michigan, to remove them from the increasing Cold War threat of Washington, DC, being attacked.

The CDSC continued teaching courses in program administration and finance, civil defense operations, and radiological monitoring among others, to State and local personnel, but by 1979, some new courses had been created on natural disaster operations. By this point in time, the FCDA came under the Department of Defense and was re-titled the Defense Civil Preparedness Agency (DCPA).

In 1979, then President Jimmy Carter brought together a number of Federal agencies that had involvement in disasters, including DCPA, and created a new, amalgamated organization, the Federal Emergency Management Agency

(FEMA). Also in 1979, President Carter dedicated the former St. Joseph's College, which closed with its merger of participants and faculty with Mount Saint Mary's University in Emmitsburg, Maryland, as the FEMA National Emergency Training Center (NETC). NETC then became the home for the National Fire Academy (NFA) and the renamed Staff College which become the Emergency Management Institute (EMI), to reflect its now broader training role. The move from Battle Creek, Michigan, to Emmitsburg was done in the Fall of 1980 and the first EMI class was conducted in January 1981.

EMI and NFA are managed independently with unique participant audiences and curricula for the emergency management and national fire communities. EMI and NFA have collaborated on curricula and programs since their inception, and share in the cost of operation of NETC. In 2011, EMI celebrated its 60th Anniversary and Legacy of Emergency Management Training and Education for the United States.[4]

Present day

EMI works to improve the competencies of United States officials at all levels of government to prevent, prepare for, respond to, recover from, and mitigate the potential effects of disasters and emergencies. EMI promotes integrated emergency management principles and practices through application of the NRF, NIMS, and an all-hazards approach. EMI is the lead national emergency management training, exercising, and education institution.

EMI offers a full-spectrum emergency management curriculum with more than 500 courses available to the integrated emergency management community, which includes: FEMA staff and disaster employees; Federal partners; State, Tribal, and local emergency managers; volunteer organizations; and first responders from across the Nation. EMI supports international emergency management with more than 50 countries participating in EMI's training and educational activities through the years, both in residence and through internationally deployed training teams.

EMI also enjoys close relations with several nationally recognized professional emergency management and related organizations and has interfaced with them through training, conferences, and exercises. Some of these significant organizations include the International Association of Emergency Managers (IAEM), National Emergency Management Association (NEMA), Association of State Floodplain Managers (ASFPM), American Public Works Association (APWA), American Society of Civil Engineers (ASCE), and American Society of Engineering Management (ASEM). In 1997, EMI was awarded the W. Edwards Deming Outstanding Training Award by the United States Department of Agriculture Graduate School at the Excellence in Government Conference. This annual award is presented to an organization for an impressive workforce development and training initiative that has measurably improved their organization's performance. EMI has provided technical support to dozens of other Federal government agencies and State offices on advanced distributed learning technology development and application.

A vital asset to FEMA's disaster operations is the Disaster Field Training Operations (DFTO) implemented by EMI. In 2010 alone, the DFTO trained 31,834 disaster response and recovery employees at disaster sites throughout the United States. EMI conducts three national-level conferences. The Institute hosts the National Preparedness Annual Training and Exercise Conference which is attended by Regional Training Managers, State Training Officers and Exercise Training Officers, State Administrative Authority Officials, and subject matter experts from a broad sector of the preparedness community. The EMI Higher Education Conference is held the first week in June for more than 400 college and university officials with current or developing programs in emergency management and hosts up to 70 separate discussion groups. The Dam Safety Conference held in February is attended by dam safety officials, hydrologists, engineers, and reclamation officials.[5]

2.6.3 Training courses overview

EMI provides training to First Responders and federal, state, local, and tribal government officials

To take an EMI course, applicants must meet the selection criteria and prerequisites specified for each course. Participants may not take the same course more than once. Instruction focuses on the four phases of emergency management: mitigation, preparedness, response, and recovery. EMI develops courses and administers resident and non-resident training programs in areas such as natural hazards (earthquakes, hurricanes, floods, dam safety), technological hazards (hazardous materials, terrorism, radiological incidents, chemical stockpile emergency preparedness), professional development, leadership, instructional methodology, exercise design and evaluation, information technology, public information, integrated emergency management, and train-the-trainers.

Approximately 5,500 participants attend resident courses each year while 100,000 individuals participate in non-resident programs sponsored by EMI and conducted by state emergency management agencies under cooperative agreements with FEMA. Another 150,000 individuals participate in EMI-supported exercises, and approximately 1,000 individuals participate in the Chemical Stockpile Emergency Preparedness Program (CSEPP).[6]

The *independent study* program at EMI consists of free courses offered to United States citizens in Comprehensive Emergency Management techniques.[7] Course IS-1 is entitled "Emergency Manager: An Orientation to the Position" and provides background information on FEMA and the role of emergency managers in agency and volunteer organization coordination. The EMI Independent Study (IS) Program, a Web-based distance learning program open to the public, delivered extensive online training with approximately 200 courses and trained more than 2.8 million individuals. The EMI IS Web site receives 2.5 to 3 million visitors a day.[8]

NETC Learning Resource Center

EMI students may potentially be granted access to NETC's Learning Resource Center (LRC). The LRC is home to the Nation's preeminent collection of all-hazards first responder resources. The LRC's 190,000 titles encompass the entire gamut of natural and man-made hazards from fires and hurricanes to tornadoes and flooding, from chemical and biological to radiological and nuclear hazards. The Emergency Management, Fire, and Emergency Medical Services collections in particular have the greatest depth and breadth of any in the United States. The LRC has a website that includes more than 22,000 downloadable items.[9]

On-campus enrollment process and transcript requests

Enrollment in on-campus EMI courses are generally limited to U.S. residents; however, each year a limited number of international participants are accommodated in EMI courses. To take a on-campus course at EMI, applicants must meet the selection criteria and prerequisites specified for each course. Applicants may not take the same course more than once. Applications for the Main Campus are to be submitted to the NETC Office of Admissions in Emmitsburg, Maryland.[10] EMI also provide transcripts at no cost to the student, or directly to a college/university of the student's request. The transcript request must also be sent to the NETC Office of Admissions.[11]

Regional on-campus options

FEMA Training Regions for EMI On-Campus Courses

For those who are not able to attend classes at the Main Campus in Emmitsburg, there are 10 FEMA regions in which EMI on-campus courses may potentially be available. The Regional Training Manager contact information is listed below.[12]

- Regional Contact Master List

- Region I, Boston, MA Serving: CT, MA, ME, NH, RI, VT
- Region II, New York, NY Serving: NJ, NY, PR, USVI
- Region III, Philadelphia, PA Serving: DC, DE, MD, PA, VA, WV
- Region IV, Atlanta, GA Serving: AL, FL, GA, KY, MS, NC, SC, TN
- Region V, Chicago, IL Serving: IL, IN, MI, MN, OH, WI
- Region VI, Denton, TX Serving: AR, LA, NM, OK, TX
- Region VII, Kansas City, MO Serving: IA, KS, MO, NE
- Region VIII, Denver, CO Serving: CO, MT, ND, SD, UT, WY
- Region IX, Oakland, CA Serving: AZ, CA, HI, NV, GU, AS, CNMI, RMI, FM
- Region IX, PAO Serving: American Samoa, CNMI,Guam, Hawaii
- Region X, Bothell, WA Serving: AK (Alaska), ID, OR, WA

Certifications

EMI offers certifications for Emergency Management Professionals

EMI offers credentials and training opportunities for United States citizens. EMI provides leadership in developing and delivering training to ensure that individuals and groups having key emergency management responsibilities at all levels of government, possess the requisite competencies to perform their jobs effectively.[13] In addition to its resident training program, EMI disseminates centrally developed training materials through a comprehensive national training program in the United States territories, and trusts. EMI has responsibility for training FEMA staff to perform their disaster response functions. Note that students do not

EMI provides training for CERT.

EMI provides training for the USCGA.

have to be employed by FEMA or be a federal employee for some of the programs.[14]

Training opportunities and certifications:

- Resident Courses

- Distance Learning Courses

- Emergency Management Professional Program (EMPP)

- Master Exercise Practitioner Program (MEPP)

- EMI School and Higher Education Programs

- Integrated Emergency Management Courses (IEMC)

- Disaster Field Training Operations (DFTO)

- Non-Resident Courses (EMI Courses Conducted by States)

- National Training, Education, and Exercise Symposium

- EMI Tribal Curriculum

- Master Trainer Program (MTP)

- Critical Infrastructure Security and Resilience

- Virtual Table Top Exercise (VTTX) Series

- Continuity Excellence Series, Level I and II

- Professional Development Series (PDS) Credential

- Advance Professional Series (APS) Credential

Emergency Management Higher Education Program

EMI in 1994 developed the Emergency Management Higher Education Program with the aim of promoting college-based emergency management higher education for future emergency managers and other interested personnel. The program works with colleges and universities, emergency management professionals, and stakeholder organizations to help create an emergency management system of sustained, replicable capability and disaster loss reduction through formal education, experiential learning, practice, and experience centered on mitigation, preparedness, response and recovery from the full range of natural, technological and intentional hazards which confront communities, States and the Nation.[15]

Transferring courses for college credits   EMI maintains a strategic partnerships with Frederick Community College. FCC has a contract with the Emergency Management Institute to provide college credit for the Independent Study (IS) Program.[16] In addition Clackamas Community College was approved by the Oregon Department of Education to accept credit for EMI coursework.[17] In addition, the University of North Texas also has a similar program[18] along with Charter Oak State College and Excelsior College.[19]

Training for the CERT and Citizen Corps

The Community Emergency Response Team (CERT) Program educates people about disaster preparedness for hazards that may impact their area and trains them in basic disaster response skills, such as fire safety, light search and rescue, team organization, and disaster medical operations. Using the EMI training learned in the classroom and during exercises, CERT members can assist others in their neighborhood or workplace following an event when professional responders are not immediately available to help.[20] In addition, EMI provides training courses for the Citizen Corps as well.[21]

Training for the Coast Guard Auxiliary

The United States Coast Guard Auxiliary requires auxiliarists to take mandatory Incident Command System courses through the Emergency Management Institute. Failure to complete the training may make them ineligible to participate in Coast Guard Auxiliary exercises, drills, or response events.[22] Auxiliarists are expected to take EMI courses that will help them to understand the Incident Command System's organization, basic terminology and common responsibilities. They are required to acquire the skills necessary to perform in an ICS support role.[23] Officers, certified coxswains, pilots, or those in a leadership role may need to take additional EMI courses pertaining to the National Incident Management System and/or the National Response Framework.[24]

2.6.4 Supervision of Emergency Management Institute

Since 1979 EMI has been under the Federal Emergency Management Agency, and had previously been under the United States Department of Defense. On April 25, 2012 Tony Russell was selected to become the Superintendent. Previously Russell served as the FEMA Region VI Administrator from December 2009 – April 2012 and he had been responsible for the oversight of FEMA operations in Arkansas, Louisiana, Oklahoma, New Mexico, and Texas. In addition, he previously served as the Acting Director of FEMA's Louisiana Recovery Office (LRO) and as a Federal Coordinating Officer (FCO) for FEMA Region VIII.[25]

2.6.5 National Emergency Management Center campus

In total the NETC campus encompasses over 107 acres (0.43 km^2), and is located in Emmitsburg, Maryland. The campus is home to many notable structures, buildings, and monuments.[26]

- Building A – Is a three-story residence hall built in 1964 and renovated in 1996. It has 96 dormitory rooms.

- Building B – The Student Center, built in 1956, is the location of a game room, pub and recreational activities. A large picture window overlooks the scenic Catoctin Mountain range.

- Building C – Built in 1956 and renovated in 1995, it has 216 dormitory rooms.

- Building D – Built in 1926 and renovated in 1965 and 1995, it is a three-story brick structure that has the

FEMA's seal before 2003.

charm of the old architecture. It consists of 39 dormitory rooms with offices and a convenience shop in the basement.

- Building E – Built in 1926 and renovated in 1966 and 1993, it is occupied by the EMI, National Fire Programs, NETC Budget offices, and computer support personnel.

- Building F – Built in 1925 and renovated in 1965 and 1995, it has 45 dormitory rooms.

- Building G – Built in 1948 and renovated in 1984 and 2001 to accommodate USFA Offices and Programs.

- Building H – Built in 1923 and renovated in 1993, it houses the NETC offices, a fully equipped Gymnasium, weight room and an indoor pool.

- Building I – Built in 1996, it serves as the Material Receipt and Distribution Center, Maintenance Facility, Management Operations and Support Services Division, Admissions Office, and O&M Support Offices.

- Building J – Built in 1966, renovated in 1993, it is the NETC classroom facility and houses the NETC staff. It includes a lobby and a tiered 249-seat auditorium.

- Building K – Built circa 1870, renovated in 1982 and 1993, it houses EMI classrooms. The three-story brick structure also contains a Dining Hall capable of seating 500 people.

- Building L – Built in 1959, renovated in 1993, it consists of 37 dormitory rooms and a conference room.

- Building M – Built in 1965, renovated in 1989, it houses two EMI classrooms and the EMI Computer Lab.

- Building N – Designed by the English-born architect, E.G. Lind, it was built in 1870 and renovated in 1987, 1992 and 2001 and is listed on the National Register of Historic Buildings

- Building O – Erected in 1839 as a chapel. The marble, alabaster altars and stained glass windows were retained when it was renovated in 1965. It was renovated again in 2006.

- Building P – The Log Cabin serves as a recreational facility overlooking peaceful Tom's Creek.

- Building Q – The Brick Barn is a service building. The ornamental brick grill windows are characteristic of the early 19th century Western Maryland construction.

- Building R – Built in 1948, renovated in 1993, located behind Building G

- Building S – Renovated in 2001 to house the NETC Joint Exercise and Simulation Lab by the NFA and EMI.

- Building T – The old Milk House of the original St. Joseph's campus, it houses Administrative Support Offices.

- Building U – A burn building complex used by the NETC for arson investigation and demonstration

- Building V – Built in 1992, it houses the Security Office

EMI main campus facilities within the NETC

-
-
-
-
-
-

National Civil Defense/Emergency Management Monument

The Monument, is the centerpiece of the EMI's main campus in the Emmitsburg. On November 13, 1999, President Bill Clinton signed a bill that granted authority to the National Civil Defense Monument Commission to construct a monument on the NETC campus. The purpose of the monument is to honor the thousands of Civil Defense and Emergency Management professionals and volunteers who have worked hard and faithfully to protect the public from both man-made and natural hazards. This monument particularly recognizes the numerous military and civilian volunteers and professionals who have gone beyond the normal call of duty to save lives and alleviate suffering in times of crises. The centerpiece of the monument is a 15-ton block of polished white Vermont granite, shaped as a three-sided pyramid, representative of the Federal, state, and local governments and their efforts in working together to accomplish a joint mission. The triangular base is 5 feet on each side, rising to 15 feet in height. The pinnacle of the monument is capped with a large, bronze American eagle, sculpted by the world-renowned sculptor, Lorenzo Ghiglieri. The base is encircled by a stone and concrete plaza with appropriately inscribed bronze state plaques embedded in concrete, surrounded by a circle of state flags. A brick wall rises approximately 3 feet in height on the back or south side of the plaza. Near the edge of the plaza are two bronze plaques bearing the names of advocates and members of the Monument Commission.[27]

2.6.6 See also

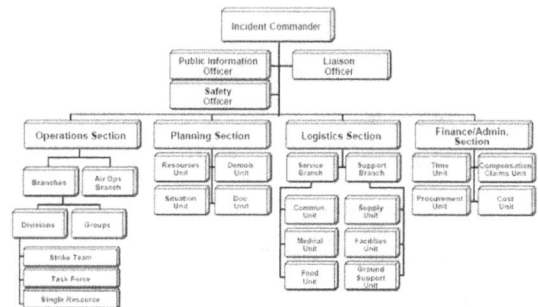

Incident Command System – Org chart

- Department of Homeland Security

- Federal Emergency Management Agency

- Incident command system

- Multiagency Coordination Systems

- National Incident Management System
- National Response Framework
- Continuity of Operations
- Emergency Management
- Community Emergency Response Team
- Incident Management Team
- Search and rescue
- FEMA photo library

2.6.7   References

[1] "Welcome to National Training and Education". Emergency Management Institute. Retrieved October 19, 2015.

[2] EMI Accredited

[3] About EMI

[4] History of EMI

[5] EMI present day info

[6] EMI Overview

[7] FEMA EMI Independent Study home

[8] IS Course Offerings

[9] About LRC

[10] Enrollment Process

[11] Requesting Transcripts

[12] http://www.fema.gov/regional-operations

[13] About EMI Training

[14] EMI Program Info

[15] About the Higher Education Program

[16] Partnership with FCC

[17] CCC Credits

[18] UNT Transfer Credit

[19] Transfer Credits by College

[20] About CERT Training

[21] EMI Supports Citizen Corps

[22] Auxiliary Requirements

[23] USCGA Training requirements

[24] USCGA Officer Requirements

[25] About the Superintendent

[26] About NETC Campus

[27] About the monument

2.6.8   External links

- Official site

Coordinates: 39°41′40″N   77°19′42″W   /   39.69444°N 77.32833°W

2.7   Executive Order 12148

Executive Order 12148 was an executive order enacted by President Jimmy Carter on July 20, 1979 to transfer and reassign duties to the newly formed agency, known as the Federal Emergency Management Agency (FEMA), created by Executive Order 12127. The order combined several federal agencies tasked with emergency preparedness and civil defense spread across the executive departments into a unified entity that was established as an independent agency, free of Cabinet interference, with authority as the lead federal agency in a presidentially-declared disaster.

The agency's place within the governmental structure was changed on March 1, 2003, when FEMA became part of the Department of Homeland Security's Emergency Preparedness and Response Directorate.

2.7.1   Revocations

E.O. 12148 also revoked the following Executive Orders or parts thereof: (E.O. Numbers)

10242; Sections 1 and 2 of 10296; 10494; 10601; 10634; 10900; 10952; 11051; 11415; 11795; 11725; and 11749.

2.7.2   External links

- National Archives page on Executive Order 12148

2.8   FEMA camps conspiracy theory

The **FEMA camps conspiracy theory** refers to the theory that the U.S Federal Emergency Management Agency (FEMA) is planning to imprison US citizens in concentration camps.[1][2][3] This is typically described as following the imposition of martial law in the United States after a major disaster or crisis. In some versions of the theory, only suspected dissidents will be imprisoned. In more extreme versions, large numbers of US citizens will be imprisoned for the purposes of extermination as a New World Order is established. The FEMA camps conspiracy theory has existed since the 1980s but it has picked up greatly in popularity since the late 2000s.[4] The theory is generally associated with the right-wing of the political spectrum.[5]

2.8.1 Arguments and variants

FEMA was established in 1979 under executive order by President Jimmy Carter. It was established to coordinate the response to a major disaster that has occurred in the United States and that overwhelms local and state authorities. However, proponents of the conspiracy theory argue that this is merely a cover for the organisation's real purpose. This is to assume control of the United States following a major disaster or threat, either a genuine one or a manufactured one. Once a disaster or threat of one comes into being, the theory goes, martial law will be declared and FEMA's emergency powers will come into operation. FEMA will then effectively be the government.[6] The constitution will be suspended and FEMA will move US citizens into specially constructed camps, many of which have already been built.[7] The organisation has been described in this context as 'the executive arm of the coming police state'.[8] Proponents of the theory often play into racial fears, asserting that FEMA will use 'urban gangs' as auxiliaries to ensure order.[9]

In many versions of the theory, 'dissidents' (typically defined as constitutionalists/patriots etc. rather than left-wingers) will merely be imprisoned.[10] Others have gone so far as to argue that they will be sent to these camps to be murdered.[11] Extreme versions of the theory state that plans are in place to imprison and kill apolitical American citizens in FEMA camps are part of a 'population control' plot.[12] FEMA conspiracies are often worked in larger conspiracy narratives about ushering in a 'New World Order', meaning a totalitarian global government.[13]

As evidence of the conspiracy theory, proponents point to supposed FEMA camps already existing in the United States. These, however, often have known, established purposes such as Amtrak facilities and Armed Forces training centers.[14] In some cases, genuine internment camps have pointed to but these have always been outside the United States.[15]

They have also cited a contingency plan (Rex 84) drafted in part by Oliver North calling for the suspension of the Constitution and the detainment of citizens in the event of a national crisis.[16] This was aimed at left-wing activists, not the Patriot types generally associated with FEMA theories.[17] This has been linked to a 1970 document by then-FEMA director Louis Guiffrida calling for the establishment of martial law in the event of an uprising by African American militants and the internment of millions of African Americans.[18]

Conspiracy theorists have used the actual internment of Japanese Americans during World War II in specifically constructed camps as evidence that such a scenario at least has historic precedent.[19] Similarly, the forced removal of Native Americans from their lands throughout US history has been pointed to.

2.8.2 History

One of the first known references to FEMA concentration camps comes from a newsletter issued by Posse Comitatus in 1982, with the warning that 'hardcore Patriots' were to be detained in them.[20] The prevalence of the conspiracy increased in line with the rise of the militia movement in the 1990s.[21] A supposed FEMA camp was featured in Linda Thompson's influential film *America Under Siege* (in reality, the 'FEMA camp' was an Amtrak repair facility). Following the 1995 Oklahoma City bombing the conspiracy theory was discussed by the Senate Judiciary Subcommittee on Domestic Terrorism.[22] The theory's inclusion in the plot of the 1998 X-Files movie showed its growing reach.[23]

Fears of FEMA declined in the early 2000s as foreign terrorists were perceived as the major threat. However, the late-2000s recession and the 2008 election of Barack Obama has renewed opposition to the federal government. In this context there has been a resurgence in the militia movement and, with it, the FEMA camps conspiracy theory.[24] This time, however, the theory has been able to reach more mainstream right-wing circles while it had previously been confined to the far-right. FOX News personality Glenn Beck, for example, devoted airtime to it on three shows, saying that he could not debunk it (although he later stated that he did not believe the theory).[25] Emails from the magazine *National Review* have also promoted the theory.[26] Sitting Congresswoman Michele Bachmann alluded to the theory while in office,[27] as have other Republican Party politicians.[28]

Such has been the upsurge that FEMA itself has gone on record saying that it has no plans to detain citizens.[29] However, in an internal memo FEMA conceded that it could not hope that convince a large number that it had no sinister plans and cautioned that it was 'better not to enter into debate on the subject.'[30] The magazine *Popular Mechanics* has published debunks of the various claims of the conspiracy theorists.[31] The Southern Poverty Law Center also points out,

'Ultimately, belief in FEMA detention camps requires one to conclude that nobody has ever escaped from one and told their story. It means believing that not one camp worker has breathed a word about his or her job. It requires assuming that not one of America's 100 senators or 435 congressmen knows of the camps or, if they do, none is alarmed enough to call for hearings. It means believing that not a single ambitious journalist connected to a national media outlet has delved into this dastardly plan. And it requires one to assume that such innocuous things as the "FEMA Trucks" signs at the Maxwell AFB — in plain view of thousands of motorists — actually betray a terrible secret.'[32]

2.8.3   References

[1] Larry Keller (2010). 'Fear of Fema' . Southern Poverty Law Center. Retrieved 10/19/2015.

[2] Jon E. Lewis (2008). *The Mammoth Book of Conspiracies.* Robinson.

[3] Mark Potok (2014). 'National Review, in e-mail blasts, warns of FEMA Camps' . Retrieved 10/19/2015.

[4] Keller (2010).

[5] Potok (2014).

[6] D.J Mulloy 'Federal Emergency Management Agency' pp. 250-251, IN: Peter Knight (ed.) (2003) *Conspiracy Theories in American History: An Encyclopedia.* ABC-CLIO

[7] Keller (2010). Mulloy (2013). p. 251.

[8] Keller (2010).

[9] Alexander Zaitchik (2010). ' 'Patriot Paranoia: A look at the top ten conspiracy theories' . Retrieved: 10/19/2015.

[10] Zaitchik (2010).

[11] Lewis (2008).

[12] Lewis (2008).

[13] Daniel Pipes (1997). *Conspiracy: How the Paranoid Style Flourishes and Where it Comes From.* Simon & Schuster. p. 8.

[14] *Popular Mechanics* (2008). 'The Evidence: Debunking FEMA Camp Myths'. . Retrieved: 10/19/2015.

[15] *Popular Mechanics (2008).*

[16] Keller (2010).

[17] Political Research Associates. *The Right-Wing Roots of Sheehan's "Secret Team" Theory* . .

[18] Keller (2010).

[19] Keller (2010).

[20] Keller (2010).

[21] Keller (2010).

[22] Zaitchik (2010).

[23] Knight (2003). P. 251.

[24] Zaitchik (2010).

[25] Keller (2010).

[26] Potok (2014).

[27] John Amato (2009). 'Michele Bachmann warns of politically correct re-education camps for young people.' . Retrieved: 10/19/2015.

[28] David Montgomery (2014). 'Candidate Hubbel wants to keep feds at bay.' . *Argus Leader.* Retrieved: 10/19/2015.

[29] Keller (2010).

[30] Zaitchik (2010).

[31] *Popular Mechanics (2008).*

[32] Keller (2010).

2.9   FEMA Photo Library

The **FEMA Photo Library** is an online gallery of photos compiled by the Federal Emergency Management Agency (FEMA) of the United States, containing more than 37,000 disaster related photographs taken since 1980. The majority of the collection is of declared disasters and there are also photographs from significant public events that have occurred on or near the National Mall in Washington, D.C. Since August 30, 2005, 6,098 images have been added to the collection; Hurricane Katrina has the most photographs in the collection with around 3,000 images.

The photographs are of hurricanes, tornadoes, floods, typhoons, fires, avalanches, ice storms, blizzards, terrorist attacks, earthquakes, and the Space Shuttle Columbia disaster.

The subjects of the photographs in the collection:

- FEMA and other disaster workers (Urban Search and Rescue, Disaster Medical Assistance Teams, the National Guard, Red Cross, US Army Corps of Engineers, US Forest Service, Internal Revenue Service, and State disaster workers) searching for and/or helping disaster victims.

- Photographs of damage to private property and public infrastructure.

- Photographs of FEMA success stories where a modification to a building or house lessened the damage from a natural disaster.

- Photographs of training.

- Photographs of FEMA provided housing for displaced disaster victims.

- Photographs of elected or appointed officials surveying the disaster and providing support to disaster workers and victims.

These photographs are in the public domain and are not copyrighted, and the collection is added to during declared disasters when multiple additions occur daily.

2.9.1 External links

- The FEMA photo library

2.10 FEMA Public Assistance

The Federal Emergency Management Agency's (FEMA) Public Assistance Program provides aid in the wake of a major disaster to state and local governments, and to certain non-profits, to help communities in their recovery efforts.[1]

The Public Assistance Program provides federal disaster grant assistance for debris removal, emergency protective measures, and the repair, replacement, or restoration of disaster-damaged property. The Public Assistance Program is meant to supplement any federal disaster grant assistance that a business or organization has already received. The Public Assistance, or PA Program, is based on a partnership between FEMA, State, and local officials. The federal share of assistance should be less than 75% of the eligible cost of emergency efforts and restoration. The remaining funds are generally allocated by the state and are distributed amongst eligible applicants.[2]

2.10.1 Applicant eligibility criteria

In order to receive a Public Assistance Grant, the applicant must register within sixty days of the disaster. The applicant must first be deemed eligible to apply for FEMA Public Assistance. Those eligible include: state government agencies, local governments, federally recognized Indian tribes, and private non-profit organizations.[3]

2.10.2 Facility eligibility criteria

For a facility to be eligible for FEMA Public Assistance, it must be located in a designated disaster area and be under the legal responsibility of an eligible applicant. The facility should have been in active use at the time of the disaster; and open to the general public.

2.10.3 Types of work covered

The FEMA Public Assistance Grant Program can only be applied to two types of disaster recovery work. The first is emergency work – this includes the debris removal and the preventative measures taken to secure the property and prevent further damage to the property and to public health. The second is permanent work – which covers the measures needed to restore, or replace, the property.

2.10.4 Grant application process

The application processes is somewhat lengthy, and can include the following steps:[4]

- A preliminary damage assessment from which an immediate needs funding and expedited payments are derived

- Applicants' Briefing where applicants receive and complete a Request for Public Assistance form

- In the event of a successful Request for Public Assistance, the applicant is assigned a public assistance coordinator

- The next phase is an introductory meeting, which is composed of the applicant and their public assistance coordinator

- The applicant's specific needs will be identified and cost estimates will be derived through the project formulation process

- Cost estimates for small projects that have been previously prepared are confirmed through a standardized validation process

- And finally if eligible, FEMA approves and processes funding for the disaster recovery project

2.10.5 References

[1] FEMA Public Assistance Public Assistance Grant Program

[2] Hawaii State Civil Defense Public Assistance

[3] Disaster Recovery Today Determining Eligibility

[4] FEMA Public Assistance Application Process

A FEMA trailer

2.11   FEMA trailer

The term **FEMA trailer**,[1][2] or **FEMA travel trailer**, is the name commonly given by the United States Government[3] to forms of temporary manufactured housing assigned to the victims of natural disaster by the Federal Emergency Management Agency (FEMA). Such trailers are intended to provide intermediate term shelter, functioning longer than tents which are often used for short-term shelter immediately following a disaster. FEMA trailers serve a similar function to the "earthquake shacks" erected to provide interim housing after the 1906 San Francisco earthquake.[4]:55[5][6]

FEMA trailers were used to house thousands of people in South Florida displaced by Hurricane Andrew in August 1992, some for as long as two and a half years.[7] After Hurricane Charley in 2004, 17,000 FEMA-issued trailers and mobile homes were successfully deployed.[8] At least 145,000 trailers were bought by FEMA to house survivors who lost their homes during the 2005 Atlantic hurricane season due to Hurricane Katrina and Hurricane Rita.[9] FEMA trailers were also made available after extensive flooding in parts of New York, Pennsylvania, and New Jersey due to Superstorm Sandy in 2012.[10][11]

News reports of health issues relating to Katrina-issue FEMA trailers began to appear in July 2006.[12] A federal report in July 2008 identified toxic levels of formaldehyde in 42% of the trailers examined, attributing problems to poor construction and substandard building materials.[13] As of 2012, two class-action lawsuits were settled, between residents of Louisiana, Mississippi, Alabama and Texas, and (1) manufacturers who built mobile homes for the Federal Emergency Management Agency (FEMA) and (2) FEMA contractors who installed and maintained them.[14]

FEMA trailers are the property of the U.S. Government and are expected to be returned after use. In 1995, some Florida residents who had difficulty finding accommodation in the aftermath of Hurricane Andrew "bought their FEMA trailers for an average of $1,100 each."[7] Following Hurricanes Katrina and Rita, the U.S. Government was left with large numbers of FEMA trailers. Surplus FEMA trailers were sold via online public auctions conducted by the General Services Administration (*see: GSA website*). The distribution and resale of Katrina FEMA trailers has been heavily criticized given the possible health risks involved.[15]

2.11.1   Description

*A **FEMA trailer** (travel trailer) in front of a formerly flooded house in New Orleans.*

Although several types and sizes of manufactured structures have been installed throughout the Gulf Coast region, most are mass-produced, one-bedroom travel trailers. These typical FEMA trailers are designed to accommodate two adults and two children. There are larger trailers and other manufactured structures that can accommodate larger families.

A typical FEMA trailer can measure 14' by 22' (308 sq. ft.)[4]:56 or 8' by 32' (256 sq. ft.).[16]:83 It consists of a master bedroom with a standard size bed, a living area with kitchen and stove, bunk beds, and a bathroom with shower. Each trailer is equipped with electricity, air conditioning, indoor heating, running cold and hot water, a propane-operated stove and oven, a small microwave oven, a large refrigerator, and a few pieces of furniture attached to the floor; usually a sofabed, a small table, and two chairs. There are only a handful of FEMA trailer designs, so nearly all trailers have the same general layout. Furniture is attached to the trailer; it is not possible to move it, and it would be illegal to do so. It is also illegal to paint the inside or outside of the trailer.[16]:82-86 Trailers have little storage space, can be very cramped, and offer little or no privacy.[17][16][18]

Each trailer is elevated about two feet (0.6 m) above the ground, on concrete supports. There is only one door on the side of each trailer, which is accessible through a wooden or aluminum stairwell. There are also long ramps for wheelchair-using occupants. Electrical service to the

FEMA trailers is installed by the local power company. (For example, in most of the Gulf Coast region power is provided by the Entergy Corporation.) Each trailer has its own power meter. Trailers have ports for telephone access, cable, and Internet access. However, these services are not handled by FEMA, and a trailer occupant must arrange to have these services installed by a local provider.[4]

The typical FEMA trailer has two propane tanks on the front of the trailer behind the master bedroom, which provide the hot water, indoor heating, and gas for the stove and oven. Running water for the trailer is usually provided by some sort of water source on the property, usually through a garden hose. Sewage is piped directly to an underground sewage main on the property. Most trailers have several windows which can be opened, as well as small light fixtures in each room.[4]

FEMA trailers are manufactured from plastic, aluminum, and particle board, and are therefore somewhat flimsy and require more maintenance than a permanent structure. They are also poorly insulated, offer little sound insulation, and are known to sway in high winds.[19]

While occupying FEMA travel trailers or mobile homes, residents are responsible for maintaining their trailers, such as keeping the trailers clean, changing lightbulbs and smoke-detector batteries, and making sure propane fuel tanks are refilled with fuel.[1]

Travel trailers and mobile homes are to be inspected once a month for the occupant's safety and convenience: if a travel trailer or mobile home requires maintenance beyond basic upkeep, residents are told to call the appropriate travel trailer maintenance hotline for their parish or county.[1] In Houston, 1200 of the 4600 trailers initially issued after Hurricane Rita required serious repairs by late 2006.[20]

2.11.2 Need

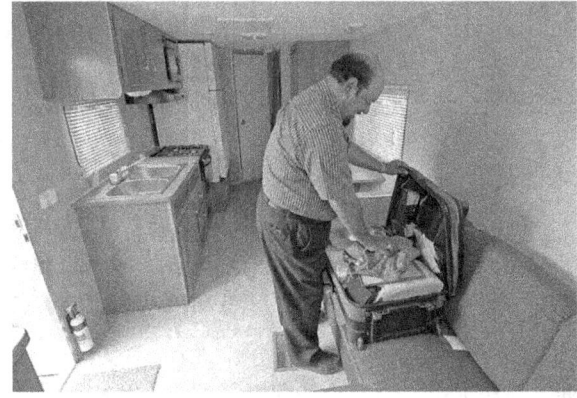

A Southern University at New Orleans professor moves into a FEMA trailer in April 2006, more than half a year after Katrina.

FEMA trailers are intended to provide temporary housing for homeowners after a disaster, until they can repair or rebuild their homes.[2] Hurricanes, tornadoes, floods and other natural disasters can cause extensive damage to residential neighborhoods, as occurred in 2005 because of Hurricane Katrina.[21][22][23]

Damaged areas may take substantial periods of time to repair. Residents may be unable to return home for long periods of time while local government officials attempt to restore basic infrastructure for water and electricity.[24] Flood damage to existing homes and apartments may require the complete removal and replacement of carpeting, flooring, insulation, and sheetrock.[25] Flood damage beyond a few inches may also destroy furniture, appliances, and other personal belongings. Buildings that have sustained significant water damage, including apartment complexes, often require extensive rebuilding and a mold-removal process known as "mold remediation" before they can be rendered safe enough for habitation.[26] Roofs may also need to be replaced or repaired.

Widespread damage in an area may cause extreme housing shortages. The extent of the rebuilding effort in such an area may cause a shortage of building contractors and materials throughout the region. This further delays the construction of new housing, and increases a need for existing apartments or motels to house incoming construction and service workers. Leasing rates for apartments may become prohibitively high for those who have lost their homes, particularly working class storm victims.[27][28] FEMA trailers, which are offered rent-free, may be the only form of habitable dwellings available to those within a disaster area. In 2005, "the devastation of housing in New Orleans and surrounding communities was so widespread that large numbers of the displaced had few options but to take up residence in FEMA parks" following extreme flood damage.[29] FEMA trailer occupants reported that they preferred the trailers to living in cars, tents, partially gutted homes, and the crowded homes of relatives.[2][30]

FEMA policy generally allows residents to live in a FEMA trailer for a period of 18 months, beginning at the time at which they receive access to the trailer. However, this period has sometimes been extended when availability of housing continued to be a serious problem for a longer time.[2] In the case of Hurricane Katrina in August 2005, deadlines were extended to allow people to live in trailers for up to 45 months.[31]

2.11.3 Application process

People within a disaster area are eligible to apply for various forms of "Housing assistance" from the Federal Emergency Management Agency (FEMA). These may include:

(1) reimbursement for short term hotel expenses (2) money or vouchers towards rental of a place to live for up to 18 months while your home is being repaired (3) money to make repairs to your home (4) money to purchase a new home if your home cannot be repaired (5) a temporary "FEMA-owned manufactured housing unit" as a last resort if no other housing options are available.[32]

To obtain disaster manufactured housing, someone must complete a FEMA application, after which they will be interviewed by a FEMA adjuster, who is similar to an insurance claims adjuster. The adjuster will determine if the damage to the home warrants temporary housing until the home is repaired.[33] After approval, the applicant is placed on a waiting list.

2.11.4 Installation

FEMA trailer (at left) alongside a Katrina-damaged house in St. Bernard Parish, Louisiana

FEMA trailer park, in what had been a neighborhood playground.

FEMA requires that a property must have running water and power BEFORE a trailer can be installed. Utilities must be available at the site: FEMA will not install trailers in neighborhoods that have no access to running water or electricity.[34]

FEMA subcontracts the installation of FEMA trailers to numerous private contractors. First, a subcontractor installs the trailer itself. After this, other contractors install the access stairs or ramps, furniture, home appliances, and water. Next, the power company must be contacted to install a power line and power meter for the trailer. Finally, a FEMA inspector will inspect the trailer for safety compliance. Only after this lengthy process, will the occupant receive the keys for their trailer.[35]

Many FEMA trailers are installed on the private property of homeowners, usually on lawns and sometimes in driveways next to the house. However, there are also numerous FEMA operated trailer parks where many storm victims have been living.[36][37]

The trailer parks operated by FEMA range from small lots, consisting of a dozen trailers in the parking lots of office buildings and supermarkets, to several massive parks occupying large plots of land with hundreds of trailers. In some cases larger parks are surrounded by a chain-link fence and brightly lit. FEMA has also provided police security and controlled access to the larger parks.[38] FEMA trailer parks have developed into small communities, with both the benefits and problems that are involved.[18][16][37][38]

2.11.5 Health problems

Reports of health problems in the FEMA trailers deployed in 2005 had begun to appear by July 2006. Residents reported breathing difficulties, persistent flu-like symptoms, eye irritation, and nosebleeds. The cause was suspected to be high formaldehyde levels in the trailers.[12] Formaldehyde was known to increase risk of cancer, asthma and other respiratory problems,[13] and children and the elderly were at greater risk.[12] As of 2008, FEMA was reported to have received 11,000 health complaints.[13]

Scientists suspected that the rush to produce large numbers of trailers might have caused manufacturers to cut corners -- using low-cost or poorly prepared materials such as glues and pressed wood in construction of the trailers, and resulting in high levels of formaldehyde emissions.[39][40] In 2007, tests on a number of FEMA trailers by the Sierra Club showed 83% had levels of formaldehyde in the indoor air at levels above the EPA recommended limit.[39][41] Congressmen Henry Waxman and Charlie Melancon requested that FEMA test the trailers and address the issue.[42]

The United States Centers For Disease Control and Prevention (CDC) performed indoor air quality testing for formaldehyde in some of the units. On February 14, 2008, the CDC published a preliminary report confirming that potentially hazardous levels of formaldehyde were found in many of the travel trailers and manufactured homes provided by FEMA.[43] In July 2008, researchers conducting a federally funded analysis reported that the toxic levels of formaldehyde were found in 42% of the trailers tested, and that they were attributable to faulty construction practices and the use of substandard building materials.[13]

As a result, FEMA modified its specifications for emergency housing, requiring that units be larger and better ventilated.[15] Construction of "manufactured homes" provided in New Jersey in 2012 was regulated by the U.S. Department of Housing and Urban Development.[44] Researchers are also investigating possibilities for better disaster housing alternatives.[45][46]

2.11.6 Distribution and resale

On March 25, 2006, FEMA issued a news release requesting residents to call the FEMA Trailer Hotline to schedule removal of unneeded FEMA trailers after Katrina and Rita-related use.[47]

In May 2009, FEMA sent eviction notices to residents in Mississippi who were living in FEMA trailers, sparking protests that the lack of available housing would result in them facing homelessness.[48] Awareness of health concerns did not deter some residents, one of whom commented that "A dry roof over a toxic trailer beats no roof at all."[49] On May 30, FEMA extended a deadline that would have evicted people from FEMA trailers. A FEMA spokesman said the organization was working with federal, state and local partners to help the residents get long-term housing.[49]

On June 3, 2009, FEMA announced plans to virtually give away roughly 1,800 mobile homes to 3,400 families displaced by Hurricane Katrina who were living in government-provided housing along the Gulf Coast.[31]

Sticker reads "NOT TO BE USED FOR HOUSING"

In 2010, the General Services Administration began holding mass public auctions of returned FEMA trailers. Units were marked with stickers that identified them as not suitable for housing, and buyers were required to sign a waiver agreeing that trailers would not be used as housing, and that new owners would be informed of the risks if the units were resold. The stickers are easy to remove, and Katrina FEMA trailers have been widely resold without any warning of possible health hazards. As many as 130,000 trailers are reported to have gone into the resale market, often in disaster zones and other places where there is strong demand for low-cost housing.[15]

In Alabama and other parts of the South, after tornadoes caused widespread destruction in 2011, post-Katrina trailers were being sold for between $2,000 and $4,000 each.[15] FEMA trailers have been sold to oil-field workers in areas such as North Dakota, where housing is in short supply.[50] Trailers were also distributed to Native American reservations,[51][52] including the Mescalero Apache in New Mexico, Standing Rock Sioux in North Dakota and South Dakota, the Cherokee in Oklahoma and the Oglala Sioux in South Dakota.[53]

It is possible to determine whether a trailer was originally one of the Hurricane Katrina trailers by checking the vehicle's VIN Number.[54] A number of resources have been suggested for those concerned about a FEMA trailer.[55]

2.11.7 Class-action settlements

As of 2012, U.S. District Judge Kurt D. Engelhardt of New Orleans approved a $42.6 million class-action lawsuit settlement. Plaintiffs, who included roughly 55,000 residents of Louisiana, Mississippi, Alabama and Texas, had alleged that the FEMA trailers emitted hazardous levels of the toxic chemical formaldehyde. The defendants included two dozen manufacturers who built mobile homes for the Federal Emergency Management Agency (FEMA), including Gulf Stream Coach Inc., Forest River Inc., Vanguard LLC and Monaco Coach Corp. A separate $5.1 million settlement dealt with claims against FEMA contractors including Shaw Environmental Inc., Bechtel Corp., Fluor Enterprises Inc. and CH2M Hill Constructors Inc., who were responsible for installing and maintaining the units.[14]

2.11.8 References

[1] "FEMA-1603-417: FEMA: Important Phone Numbers for FEMA Travel Trailer Occupants". *FEMA.gov*. Federal Emergency Management Agency. March 25, 2006.

[2] Stuckey, Mike (October 25, 2005). "New life in a FEMA trailer". *MSNBC News* (Rising from Ruin). Retrieved 11 September 2015.

[3] "Safety Precautions Advised For Fema Travel Trailer Residents". *FEMA.gov*. Federal Emergency Management Agency. December 11, 2006. Retrieved 11 September 2015.

[4] Brown, Joseph Darnell (August 3, 2013). *A Minimalistic Approach to Adaptive , Emergency Relief S tructures Embodied by Promoting a Downsized Way of Life* (PDF). Auburn, Alabama: Auburn University (Thesis).

[5] Rafkin, Louise (February 4, 2012). "Earthquake Refugee Cottages". *The Texas Tribune*. Retrieved 11 September 2015.

[6] "1906 Earthquake Refugee Shacks". *The Western Neighborhoods Project*. Retrieved 11 September 2015.

[7] Navarro, Mireya (February 27, 1995). "New Housing for Hurricane's Last Victims". *The New York Times*. p. A10. Retrieved 11 September 2015.

[8] "Release LTR-06-044: FEMA Concludes 2004 Housing Transition Effort In Florida". *Federal Emergency Management Agency*. November 2, 2006.

[9] Watson, Bruce (August 28, 2010). "The Awful Odyssey of FEMA's Hurricane Katrina Trailers". *Daily Finance*. AOL.com. Retrieved 11 September 2015.

[10] Pearson, Erica; Durkin, Erin; Connor, Tracy (November 5, 2012). "Feds may put up FEMA trailers in New York to house tens of thousands whose homes were devastated in superstorm Sandy". *New York Daily News*. Retrieved 14 September 2015.

[11] Haddon, Heather (December 12, 2012). "New York City Shuns Trailers Welcomed in N.J.". *The Wall Street Journal*. Retrieved 14 September 2015.

[12] Brunker, Mike (July 25, 2006). "Are FEMA trailers 'toxic tin cans'? Private testing finds high levels of formaldehyde; residents report illnesses". *MSNBC.com*. Retrieved 11 September 2015.

[13] Hsu, Spencer S. (2008-07-03). "Toxicity in FEMA Trailers Blamed on Cheap Materials, Low Construction Standards". *Washington Post*.

[14] Brunker, Mike (September 28, 2012). "Class-action suit against FEMA trailer manufacturers settled for $42.6 million". *NBC News*. Retrieved 11 September 2015.

[15] Cohen, Ariella (May 19, 2011). "Despite health fears, trailers are housing disaster victims". *The Lens*. Retrieved 14 September 2015.

[16] Browne, Katherine E. (2015). *Standing in the Need Culture, Comfort, and Coming Home after Katrina*. University of Texas Press.

[17] Kunzelman, Michael (March 26, 2006). "A new normal' life for family sharing FEMA trailers". *Northeast Mississippi Daily Journal*. Retrieved 17 September 2015.

[18] Peacock, Walter Gillis, ed. (1997). *Hurricane Andrew : ethnicity, gender and the sociology of disasters* (1st ed.). London: Routledge. pp. 129–131. ISBN 978-0415168113. Retrieved 17 September 2015.

[19] Reeder, Linda. "Enough Already: Katrina, FEMA Trailers, and the Road Forward". *The American Institute of Architects*.

[20] "Users damaged 1,200 FEMA trailers since Rita, agency says". *The Houstonian*. November 2, 2006.

[21] Knabb, Richard D.; Rhome, Jamie R.; Brown, Daniel P. (10 August 2006). *Tropical Cyclone Report Hurricane Katrina 23 - 30 August 2005 (updated 10 August 2006)* (PDF). National Hurricane Center. Retrieved 11 September 2015.

[22] US Department of Commerce, "Service Assessment: Hurricane Katrina August 23–31, 2005" (June 2006), pp. 10/16, NOAA's National Weather Service, Silver Spring, MD, webpage: NWS-PDF: page 7 (surge 26-28 feet, 9 m), page 50: "Appendix C: Tornado Reports Associated with Hurricane Katrina" (62 tornadoes).

[23] Gary Tuchman, Transcript of "Anderson Cooper 360 Degrees" (2006-08-29) 19:00 ET, *CNN*, CNN.com webpage: CNN-ACooper082906: GARY TUCHMAN, CNN Correspondent: Responds to Anderson Cooper that it felt like it would never end, saying winds were at least 100 miles per hour in Gulfport for seven hours, between about 7:00 a.m. and 2:00 p.m. For another five or six hours, on each side of that, they [Gulfport] had hurricane-force winds over 75 miles per hour; much of the city [Gulfport, Mississippi, in Harrison County] of 71,000 was then under water.

[24] Becker, Christine (2009). "Disaster recovery: a local government responsibility". *PM Magazine* **91** (2). Retrieved 17 September 2015.

[25] "Repairing Your Flooded Home" (PDF). *American Red Cross*. Retrieved 17 September 2015.

[26] "Mold Remediation". *FEMA.gov*. Retrieved 17 September 2015.

[27] Chang, Yan; Wilkinson, Suzanne; Potangaroa, Regan; Seville, Erica (June 2010). "Resourcing challenges for post-disaster housing reconstruction: a comparative analysis". *Building Research & Information* **38** (3): 247–264. doi:10.1080/09613211003693945. Retrieved 17 September 2015.

[28] Esnard, Ann-Margaret; Sapat, Alka (2014). *Displaced by disaster : recovery and resilience in a globalizing world* (PDF). ISBN 978-0415856041.

[29] Lee, Matthew R.; Weil, Frederick D.; Shihadeh, Edward S. (2007). "The FEMA Trailer Parks: Negative Perceptions and the Social Structure of Avoidance" (PDF). *Sociological Spectrum* **27**: 741–766. doi:10.1080/02732170701534242. Retrieved 17 September 2015.

[30] Martinez, Humberto (December 20, 2008). "FEMA trailer Christmas keeps family's spirits bright". *Beaumont Enterprise*. Retrieved 17 September 2015.

[31] "Katrina Trailers for Sale -- for $5 or Less: White House to sell homes to hurricane-displaced families for as little as $1". *MSNBC.com*. June 3, 2009.

[32] "What Specific Items are Covered by "Housing Assistance"?". *FEMA.org*. Retrieved 17 September 2015.

[33] "FEMA Information". *Tulane University*. Retrieved 17 September 2015.

[34] "How to get a FEMA trailer (be prepared to wait)". *Daily Comet*. October 21, 2005. Retrieved 17 September 2015.

[35] "FEMA Trailers". *Entergy News*. Retrieved 17 September 2015.

[36] McCarthy, Francis X. (August 27, 2010). *FEMA Disaster Housing: From Sheltering to Permanent Housing* (PDF). Washington, D.C.: Congressional Research Service.

[37] The Louisiana Justice Institute (2008). *NO WAY TO TREAT OUR PEOPLE: FEMA Trailer Residents 30 Months after Katrina* (PDF). Retrieved 17 September 2015.

[38] Adams, Vincanne (2013). *Markets of sorrow, labors of faith : New Orleans in the wake of Katrina*. Durham and London: Duke University Press. p. 40. ISBN 978-0822354499. Retrieved 17 September 2015.

[39] Dilanian, Ken (July 19, 2007). "FEMA blasted over trailer toxins". *USA Today*. Retrieved 11 September 2015.

[40] Hsu, Spencer S. (2008-05-25). "Safety Lapses Raised Risks In Trailers for Katrina Victims". *Washington Post*. Retrieved 11 September 2015.

[41] United States (2007). *Federal housing response to Hurricane Katrina : hearing before the Committee on Financial Services, U.S. House of Representatives, One Hundred Tenth Congress, first session, February 06, 2007*. Washington: U.S. G.P.O. ISBN 9780160786358.

[42] Spake, Amanda (February 14, 2007). "Dying for a Home: Toxic Trailers Are Making Katrina Refugees Ill". *The Nation*. Retrieved 11 September 2015.

[43] Formaldehyde Levels in FEMA-Supplied Trailers (PDF)

[44] Vilensky, Mike; Maloney, Jennifer (November 16, 2012). "What's in a Name? A Lot, for FEMA's Housing Units". *The Wall Street Journal*. Retrieved 17 September 2015.

[45] "A Response to Katrina: The making of the largest, most efficient modular building order in US history..." (PDF). *Federation of American Scientists*. 24 June 2014. Retrieved 17 September 2015.

[46] Clark, Steve (June 18, 2014). "Disaster Pilot Project Seeks Better Alternative to FEMA Trailers". *The Brownsville Herald*. Retrieved 17 September 2015.

[47] "FEMA: Important Phone Numbers for FEMA Travel Trailer Occupants" (news), Federal Emergency Management Agency (uses term "FEMA Trailer Hotline"), March 25, 2006, FEMA.gov webpage: FEMA-24519.

[48] Fausset, Richard (June 21, 2006). "FEMA Calls Off Trailer Evictions". *Los Angeles Times*. Retrieved 17 September 2015.

[49] "Katrina protesters seek housing, changes". *UPI.com*. June 2, 2009. Retrieved 11 September 2015.

[50] Smith, Heather (August 27, 2015). "People are still living in FEMA's toxic Katrina trailers — and they likely have no idea". *Grist.com*. Retrieved 16 September 2015.

[51] Cooper, Kenneth J. (May 2, 2011). "As Housing Shortage Worsens, Tribes Forced to Use FEMA Trailers". *America's Wire, News Report*. Retrieved 17 September 2015.

[52] "Vacant FEMA trailers from Katrina given to Indian tribes in need of housing". *Associated Press*. July 6, 2011.

[53] "FEMA ships Katrina trailers to Native American reservations". *Politics Report*. July 13, 2011. Retrieved 14 September 2015.

[54] "Do you own a post-Katrina FEMA trailer?". *Grist.com*. Retrieved 16 September 2015.

[55] Smith, Heather (August 27, 2015). "So you're living in one of FEMA's Katrina trailers". *Grist.com*. Retrieved 16 September 2015.

2.11.9 External links

- Evans, Derrick (2007). "The KatrinaRitaVille Express Tour Activists take a FEMA trailer on the road". *Dollars & Sense* (272). Retrieved 11 September 2015.

- "FEMA Trailers For Sale". *GovernmentAuctions.org*. GovernmentAuctions.org, A Division of Cyweb Holdings, Inc. Retrieved 11 September 2015.

2.12   Flood insurance rate map

A **flood insurance rate map** (**FIRM**) is an official map of a community within the United States that displays the floodplains, more explicitly special hazard areas and risk premium zones, as delineated by the Federal Emergency Management Agency (FEMA).[1] The term is used mainly in the United States but similar maps exist in many other countries, such as Australia.

2.12.1 Uses

FIRMs display areas that fall within the 100-year flood boundary. Areas that fall within the boundary are called *special flood hazard areas* (SFHAs) and they are further divided into insurance risk zones. The term 100-year flood indicates that the area has a one-percent chance of flooding in any given year, not that a flood will occur every 100 years.[2]

Such maps are used in town planning, in the insurance industry, and by individuals who want to avoid moving into a home at risk of flooding or to know how to protect their property. FIRMs are used to set rates of insurance against risk of flood and whether buildings are insurable at all against flood. It is similar to a topographic map, but is designed to show floodplains. Towns and municipalities use FIRMs to plan zoning areas. Most places will not allow construction in a flood way.

2.12.2 Creation process

In the United States the FIRM for each town is occasionally updated. At that time a preliminary FIRM will be published, and available for public viewing and comment. FEMA sells the official FIRMs, called *community kits*, as well as an updating access service to the maps. There are also some companies that sell software to locate land parcels or real estate on digitized FIRMs. These FIRMs are very useful in identifying whether a land or building is in flood zone or not. If so, there are different zones to indicate flood effect. The different zones are zone A, AE, V, etc.

2.12.3 Conversion to DFIRMs

In 2004, FEMA began a project to update and digitize the flood plain maps at a yearly cost of $200 million. The new maps usually take around 18 months to go from a preliminary release to the final product. During that time period FEMA works with local communities to determine the final maps.[3]

2.12.4 Louisiana and FEMA

In early 2014, two congressmen from Louisiana, Bill Cassidy and Steve Scalise, asked FEMA to consider the width of drainage canals, water flow levels, drainage improvements, pumping stations and computer models when deciding the final flood insurance rate maps.[4]

2.12.5 See also

- National Flood Insurance Program
- Floodplain

2.12.6 References

[1] "Flood Insurance Rate Maps (FIRMs)". Washington, D.C.: U.S. Department of Homeland Security - Federal Emergency Management Agency. 2009-06-04. Retrieved 2010-02-16.

[2] Reilly, John W.; Spodek, Marie S. (2006), *The Language of Real Estate* (6 ed.), Chicago: Dearborn Financial Publishing, Inc., p. 188, ISBN 978-1-4195-2479-0

[3] Crumb, Michael J. (2010-01-23). "Cities Say New FEMA Flood Maps Are Full of Errors". *ABC News*. Retrieved 2010-01-28.

[4] Martin, Aaron (2014-01-06). "Cassidy, Scalise push for fair flood rate maps". *Ripon Advance* (Washington, DC). Retrieved 2014-01-10.

2.12.7 External links

- Free FIRMettes from FEMA:

2.13 Forward Challenge

Forward Challenge 06 was an exercise in crisis operations and continuity of government operation conducted by the Federal Emergency Management Agency and other agencies in June 2006. The exercise included activities at Mount Weather in Virginia. [1]

2.13.1 References

[1] Vanderbilt, Tom (August 28, 2006). "Is this Bush's secret bunker?". The Guardian. Retrieved 2009-05-23.

2.14 HAZUS

Hazus is a geographic information system-based natural hazard developed and freely distributed by the Federal Emergency Management Agency (FEMA).

In 1997 FEMA released its first edition of a commercial off-the-shelf loss and risk assessment software package built on GIS technology. This product was termed HAZUS97. The current version is Hazus-MH V2.0 where MH stands for

HAZUS-MH Logo

'Multi-Hazards' and was released in 2004. Currently, Hazus can model four types of hazards: flooding, hurricanes, coastal surge, and earthquakes. The model estimates the risk in three steps. First, it calculates the exposure for a selected area. Second, it characterizes the level or intensity of the hazard affecting the exposed area. Lastly, it uses the exposed area and the hazard to calculate the potential losses in terms of economic losses, structural damage, etc.

Although it was developed with the US continent in focus, the Hazus toolset has been adopted by emergency management organizations worldwide such as Singapore, Canada, Australia, and Pakistan.

2.14.1   Description

US nationally applicable standardized methodology that contains models for estimating potential losses from earthquakes, floods and hurricanes. Hazus uses Geographic Information Systems (GIS) technology to estimate physical, economic and social impacts of disasters. It graphically illustrates the limits of identified high-risk locations due to earthquake, hurricane and floods. Users can then visualize the spatial relationships between populations and other more permanently fixed geographic assets or resources for the specific hazard being modeled, a crucial function in the pre-disaster planning process.

Hazus is used for mitigation and recovery, as well as preparedness and response. Government planners, GIS specialists and emergency managers use Hazus to determine losses and the most beneficial mitigation approaches to take to minimize them. Hazus can be used in the assessment step in the mitigation planning process, which is the foundation for a community's long-term strategy to reduce disaster losses and break the cycle of disaster damage, reconstruction and repeated damage. Being ready will aid in recovery after a natural disaster.

As the number of Hazus users continues to increase, so do the types of uses. Increasingly, Hazus is being used by states and communities in support of risk assessments that perform economic loss scenarios for certain natural hazards and rapid needs assessments during hurricane response. Other communities are using Hazus to increase hazard awareness. Successful uses of Hazus are profiled under Mitigation and Recovery and Preparedness and Response. Emergency managers have also found these map templates helpful to support rapid impact assessment and disaster response.

2.14.2   Requirements

Although Hazus-MH itself is free, it requires the users to have ArcGIS with ArcView license level.[1] In addition, ArcGIS Spatial Analyst extension is required for Flood Model. Further more it currently is only available for use with ArcGIS version 10.0 (Current version is 10.2.2). As part of a major effort to modernize Hazus, a number of updates are in progress. In late 2014, an update was released to bring Hazus up to compatibility with ArcGIS 10.2.2 and Windows 8. Later in the Hazus Modernization process, new functional enhancements will be implemented in the flood module and the underlying code of Hazus will be re-architected to align with current practices, enabling future development.

2.14.3   Advanced analysis

Mapping results from the Advanced Engineering Building Module (AEBM) for earthquake hazards Salt Lake Community.[2]

2.14.4   Hazus User Group Community

Hazus has a substantial user group community that includes a Hazus LinkedIn group[3] and several Hazus User Groups across the nation "providing a network of HAZUS users, promoting and supporting the application of the FEMA HAZUS software for disaster mitigation, planning, response and recovery. This group is supported by HAZUS.org the independent on-line voice for the HAZUS user community and the ultimate resource for everything HAZUS".[3] Its reach includes 40 User Groups managed by 38 User Group Leaders, "Hazus User Groups (HUGs) provide a network of public and private sector organizations and industry partnerships to collaborate and disseminate Hazus information and data throughout the nation. Hazus User Group members include emergency managers, Geospatial Information System (GIS) specialists, geologists, state and local planners, consultants and other stakeholders who use Hazus software for risk assessment activities".[4] Each Hazus User Group conducts its own training, holds seminars, and occasionally holds group-wide User Group Conference calls.[5]

2.14.5   References

[1] "Requirements". *Federal Emergency Management Agency.* Retrieved 23 February 2015.

[2] "Descriptive work around". *http://code.google.com". Retrieved 24 February 2015.*

[3] "About the HAZUS Group". *LinkedIn.* Retrieved 24 February 2015.

[4] "Hazus User Groups". *Federal Emergency Management Agency.* Retrieved 24 February 2015.

[5] "Main Page". *http://www.usehazus.com/". Retrieved 24 February 2015.*

2.14.6   External links

- FEMA Hazus website

- USEHAZUS An All Hazus Web Space

- Hazus website

- Hazus LinkedIn

2.15   Main Core

Main Core is the code name of an American governmental database that is believed to have been in existence since the 1980s. It is believed that Main Core is a federal database containing personal and financial data of millions of United States citizens believed to be threats to national security.[1] The data which is believed to come from the NSA, FBI, CIA, and other sources,[1] is collected and stored without warrants or court orders.[1] The database's name derives from the fact that it contains "copies of the 'main core' or essence of each item of intelligence information on Americans produced by the FBI and the other agencies of the U.S. intelligence community."[1]

The Main Core database is alleged to have originated with the Federal Emergency Management Agency (FEMA) in 1982, following Ronald Reagan's Continuity of Operations plan outlined in the National Security Directive (NSD) 69 / National Security Decision Directive (NSDD) 55, entitled

"Enduring National Leadership," implemented on September 14, 1982.[1][2]

As of 2008 there were allegedly eight million Americans listed in the database as possible threats, often for trivial reasons, whom the government may choose to track, question, or detain in a time of crisis.[3]

The existence of the database was first asserted in May 2008 by Christopher Ketcham[4] and in July 2008 by Tim Shorrock.[2]

2.15.1   See also

- FBI Index

- *Inslaw Inc. v. United States Government*

- Investigative Data Warehouse

- National Security and Homeland Security Presidential Directive

- NSA warrantless surveillance controversy

- PRISM (surveillance program)

2.15.2   References

[1] Shorrock, Tim (July 23, 2008). "Exposing Bush's historic abuse of power". Salon.com. Retrieved 2010-12-19.

[2] Goodman, Amy (July 25, 2008). "Main Core: New Evidence Reveals Top Secret". Democracy Now. Retrieved 2010-12-19.

[3] Christopher Ketcham, Is the government compiling a secret list of citizens to detain under martial law? at the Wayback Machine (archived August 31, 2008), RADAR Online, 15 May 2008

[4] Satyam Khanna, Govt. May Have Massive Surveillance Program For Use In 'National Emergency,' 8 Million 'Potential Suspects', Think Progress blog, May 20, 2008.

2.15.3   External links

- Radar article by Christopher Ketcham, May/June 2008

- Main Core, PROMIS and the Shadow Government by Ed Encho, February, 2009

- Salon's New Revelations on Illegal Spying at Electronic Frontier Foundation

- NSA's Domestic Spying Grows As Agency Sweeps Up Data by Siobhan Gorman, Updated March 10, 2008 12:01 a.m. ET

2.16 Mount Weather Emergency Operations Center

The **Mount Weather Emergency Operations Center** is a civilian command facility in the U.S. Commonwealth of Virginia, used as the center of operations for the Federal Emergency Management Agency (FEMA). Also known as the **High Point Special Facility** (HPSF), its preferred designation since 1991 is "SF".[1]

The facility is a major relocation site for the highest level of civilian and military officials in case of national disaster, playing a major role in U.S. continuity of government (per the Continuity of Operations Plan).[2]

Mount Weather is the location of a control station for the FEMA National Radio System (FNARS), a high frequency radio system connecting most federal public safety agencies and U.S. military with most of the states.[3] FNARS allows the president to access the Emergency Alert System.[4]

The site was brought into the public eye by *The Washington Post*, when the government facility was mentioned while reporting on the December 1, 1974, crash into Mount Weather of TWA Flight 514, a Boeing 727 jetliner.[5]

2.16.1 Location

Located in the Blue Ridge Mountains,[2] access to the operations center is available via State Route 601 (also called Blueridge Mountain Road) in Bluemont, Virginia.[6] The facility is located near Berryville, 48 miles (77 km) from Washington, D.C.[7]

The site was originally opened as a weather station in the late 1800s.[8] It was used as a Civilian Public Service facility (Camp #114) during World War II.[9][10] At that time there were just two permanent buildings on the site: the administration/dormitory building, and the laboratory. Those buildings still stand, supplemented by many more modern buildings.

The underground facility within Mount Weather, designated "Area B", was completed in 1959. FEMA established training facilities on the mountain's surface ("Area A") in 1979.[11]

The above-ground portion of the FEMA complex (Area A) is at least 434 acres (176 ha). This measurement includes a training area of unspecified size.[11] Area B, the underground component, contains 600,000 square feet (56,000 m^2).[7]

2.16.2 Evacuations

According to a letter to the editor of *The Washington Post*, after the September 11 attacks, most of the congressional leadership was evacuated to Mount Weather by helicopter.[7][12][13]

Between 1979 and 1981, the National Gallery of Art developed a program to transport valuable paintings in its collection to Mount Weather via helicopter. The success of the relocation would depend upon how far in advance warning of an attack was received.[14]

2.16.3 In the media

The first video of Mount Weather shot from the air to be broadcast on national TV was filmed by ABC News producer Bill Lichtenstein, and was included in the 1983 *20/20* segment "Nuclear Preparation: Can We Survive", featuring *20/20* correspondent Tom Jarriel. Lichtenstein flew over the Mt. Weather facility with an ABC camera crew. The news magazine report also included House Majority Leader Tip O'Neill and Representative Ed Markey, confirming that there were contingency plans for the relocation of the United States government in the event of a nuclear war or major disaster.

Both Mount Weather and The Greenbrier were featured in the A&E documentary *Bunkers*. The documentary, first broadcast on October 23, 2001, features extensive interviews with engineers and political and intelligence analysts, providing rare insights into the secret installations. The documentary compared The Greenbrier and Mount Weather to Saddam Hussein's control bunker buried beneath Baghdad. The documentary features interior video of The Greenbrier as well as the Baghdad bunker, which survived direct hits from seven Joint Direct Attack Munition bombs during the Battle of Baghdad in 2003.

Author William Poundstone investigated Mount Weather in his 1989 book *Bigger Secrets*.

2.16.4 In popular culture

Mount Weather is mentioned extensively in Milton William Cooper's 1991 book *Behold A Pale Horse*.

The novel *Seven Days in May* mentions a facility called Mount Thunder, a reference to Mount Weather, but the road descriptions in the book make it quite clear that it is the same facility.[15]

A facility similar to Mount Weather is featured in the beginning of the 2002 film *The Sum of All Fears*, based on the Tom Clancy novel of the same name. The fictional

U.S. president is taken to a facility located inside Sugarloaf Mountain in Maryland during a rehearsal of emergency operation plans following a Russian nuclear attack.

Mount Weather was mentioned as the emergency facility in the case of a Soviet nuclear attack from Cuba during the Cuban Missile Crisis in the 2000 film *Thirteen Days*.

In the final episode of *The X-Files*, entitled "The Truth", ex-FBI agent Fox Mulder enters the Mount Weather complex, which is controlled by a shadow government.

In the 2008 remake of *The Day the Earth Stood Still*, Klaatu's robot is taken to Mount Weather for analysis.

In the episode of *Earth: Final Conflict* (Season 2), entitled "Message in a Bottle", Mount Weather is a hideout of a group of United States soldiers against the Taelon aliens.

In the Vince Flynn novel *Memorial Day*, the main character Mitch Rapp recovers a nuclear bomb brought to Washington, D.C., by terrorists. Unable to disarm the device, and without enough time to get it clear of the area, he transports it by helicopter to an evacuated Mount Weather where he sends it by elevator to the deepest level and seals the facility. It detonates and the blast is contained, sacrificing the facility, but saving the surrounding area.

A similar plot point takes place in Pete Earley novel "Lethal Secrets" where a Soviet thermonuclear bomb is found in the old Soviet embassy in Washington, DC. Due to a reversed altimeter, the bomb cannot be flown out of the city and dumped in the ocean, nor can it be defused due to numerous fail deadlies. The bomb is later taken to Mount Weather, taken down the elevator to the lowest level before being sealed in before detonating. The blast sacrifices the facility and is barely contained, but it saves Washington, DC and prevents World War III.

In the 2002 film *xXx* Mount Weather serves as the command center for the NSA's secretive spy program headed by Samuel L. Jackson.

In the 2012 video game *Call of Duty: Black Ops II*, Mount Weather is referenced by the automated bus driver in TranZit of the Zombies game mode. This is significant since the Zombies storyline in Black Ops II picks up after a catastrophic event ravages Earth and the world is consumed by a zombie apocalypse.

Mount Weather is the attempted landing site in the beginning of the first of the *The 100 trilogy* and a 2014 television series of which it is based on. It later becomes the setting of much of the second and third books *Day 21* and *Homecoming*, and the television series' second season.

The third book in the Lux Series; Opal, by Jennifer L. Armentrout also features Mount Weather. Two teenagers are being held captive in Mount Weather as part of an experiment. A group of teenagers creates a mission to rescue their friends.

2.16.5 See also

- Continuity of government
- Raven Rock Mountain Complex
- Warrenton Training Center

2.16.6 Notes

[1] Gup, Ted (December 9, 1991). "Civil Defense Doomsday Hideaway". *Time*: 26–30.

[2] "Fire Departments" (PDF). *The Lay of the land: the Center for Land Use Interpretation Newsletter* (Culver City, CA: The Center for Land Use Interpretation): 6–7. Spring 2002. Retrieved April 3, 2008.

[3] "Opportunities With OES ACS Program". *OES Auxiliary Communications Service Homepage*. Governor's (California, USA) Office of Emergency Services. Retrieved April 2, 2008.

[4] Merlin, Ross Z. (2004). "Communications Systems for Public Health Contingencies" (PDF). DHS/FEMA Wireless Program Management Team. Archived from the original (PDF) on June 25, 2008. Retrieved April 2, 2008.

[5] Mount Weather "High Point Special Facility (SF), Western Virginia Office" Check |url= scheme (help).

[6] Bedard, Paul (December 4, 2001). "Things That Go Bump In The Night At Cheney's Cave". *White House Weekly*. p. 1.

[7] Schwartz, Stephen I. (August 9, 2006). "Near Washington, Preparing for the Worst". *The Washington Post*. p. A16.

[8] "Mt. Weather".

[9] "CPS Camp # 114".

[10] "CPS Unit Number 114-01".

[11] McGrath, Gareth (January 30, 2002). "Training Site Bunker Used After Sept. 11 Terror Attacks". *Morning Star* (Wilmington, NC). pp. 1B, 6B.

[12] "Mount Weather". *Weapons of Mass Destruction (WMD)*. GlobalSecurity.org. April 27, 2005. Retrieved November 27, 2009.

[13] Jeanne Meserve and Mallory Simon (November 26, 2009). "Web site posts what it says are half million text messages from 9/11". *CNN* (Turner Broadcasting System, Inc). Retrieved November 27, 2009.

[14] Gup, Ted (October 10, 1992). "Grab That Leonardo!". *Time*. Retrieved April 3, 2008.

[15] Vanderbilt, Tom (August 28, 2006). "Is This Bush's Secret Bunker?". Comments and Features. *The Guardian*. p. 12. Retrieved April 2, 2008.

2.16.7 References

- Emerson, Steven (August 7, 1989). "America's Doomsday Project". *U.S. News & World Report*: 26–31.

- —— (August 10, 1992). "The Doomsday Blueprints". *Time*: 32–39.

2.16.8 External links

- 1962 Mount Weather Operating Documents

- FEMA page on its Mt. Weather operations as saved by the Internet Archive Wayback Machine on March 30, 2005

Coordinates: 39°03′47″N 77°53′20″W / 39.063°N 77.889°W

2.17 National Disaster Medical System

National Disaster Medical System logo.

The **National Disaster Medical System** (**NDMS**) is a section of the United States Department of Health and Human Services (HHS) responsible for managing Federal government's medical response to major emergencies and disasters.

The overall purpose of the NDMS is to supplement an integrated National medical response capability for assisting State and local authorities in dealing with the medical impacts of major peacetime disasters and to provide support to the military and the Department of Veterans Affairs medical systems in caring for casualties evacuated back to the U.S. from overseas armed conventional conflicts.[1]

NDMS's federal partners include the Federal Emergency Management Agency, Department of Defense (DOD), and the Department of Veterans Affairs (VA).

NDMS also interfaces with state and local Departments of Health, as well as private hospitals.

2.17.1 Organization

NDMS has three major components:[1]

1. Emergency medical response by civilian medical teams, equipment, and supplies to a disaster area when local medical resources are overwhelmed

2. Movement of ill and injured patients from a disaster area to areas unaffected by the disaster

3. Definitive care of patients at hospitals in areas unaffected by the disaster.

Over 8,000 NDMS civilian volunteer medical personnel are organized into a number of types of medical teams, designed to accomplish the emergency medical response mission.

NDMS Teams

The NDMS is made up of several smaller teams that each focus on a particular area of disaster relief.

- **Disaster Medical Assistance Team** (DMAT) - provides medical care during a disaster or other incident.[2]

- **National Medical Response Team** (NMRT) - provides mass decontamination and medical care to victims of a release of Weapons of Mass Destruction, or a large scale release of Hazardous Material.

- **Disaster Mortuary Operational Response Team** (DMORT) - provide victim identification and mortuary services during a disaster or other incident.[3]

- **National Veterinary Response Team** (NVRT) - provides assistance in assessing the need for veterinary services following major disasters or emergencies[4]

- **Federal Coordinating Centers** (FCCs) - recruit hospitals and maintain local non-Federal hospital participation in the NDMS and coordinate exercise development and emergency plans[5]

- **National Pharmacy Response Team** (NPRT) - assists in chemoprophylaxis or vaccination of large numbers of citizens in response to an emergency involving a disease outbreak[6]

- **International Medical Surgical Response Team** (IMSuRT) - widely recognized as a specialized team, trained and equipped to establish a fully capable field surgical facility anywhere in the world.[7]

- **National Nurse Response Team** (NNRT) - A specialty DMAT designed for a scenario requiring hundreds of nurses to assist in chemoprophylaxis, mass vaccination programs, or situations that overwhelm the nation's supply of nurses.[8]

- **Incident Response Coordination Team** (IRCT) - Provides the field management component of the Federal public health and medical response. The IRCT provides liaisons in the field to coordinate with jurisdictional, Tribal, or State incident management and provides the field management and coordination for deployed HHS and other ESF #8 assets to integrate those assets with the State and local response.[9]

Over 1,800 civilian hospitals in the U.S. are members of NDMS. Their role is to provide approximately 100,000 treatment beds to support NDMS operations in an emergency. When a civilian or military crisis requires the activation of the NDMS system, participating hospitals communicate their available bed space to a central control point. Patients can be distributed to a number of hospitals without overwhelming any one facility with casualties.

2.17.2   Operations

Under the NDMS, movement (evacuation) of patients from a disaster area is coordinated by the FCCs in each of the 10 FEMA regions. The actual transport is conducted by the Department of Defense. Patients arriving in a region are then dispersed to a local NDMS participating hospital.

In the aftermath of Hurricane Katrina in the fall of 2005, the NDMS system activated almost all of their civilian medical teams to assist victims in Texas, Louisiana, and Mississippi; helped evacuate hundreds of medical patients from the affected areas; and augmented medical staffing levels at hospitals impacted by the evacuations.

2.17.3   Parent Agencies

NDMS was originally under the U.S. Public Health Service (USPHS) within The Department of Health and Human Services(DHHS). In 2003, as a direct result of the September 11 attacks in 2001 the newly formed Department of Homeland Security (DHS), requested, and was granted convening authority over NDMS which was then placed under the direction of the Federal Emergency Management Agency (FEMA).

After Hurricane Katrina, amidst allegations of mismanagement (etc.), NDMS was removed from FEMA and sent back to DHHS, as legislated by an Act of Congress entitled. "the Pandemic and All Hazards Preparedness Act (PAHPA)", (Public Law 109-417), effective January 1, 2007.

This positioned NDMS, organizationally, within the Assistant Secretary for Preparedness and Response (ASPR) Office of Preparedness and Emergency Operations (OPEO).

The OPEO is responsible for developing operational plans, analytical products, and training exercises to ensure the preparedness of the Office, the Department, the Federal Government and the public to respond to domestic and international public health and medical threats and emergencies. OPEO is also responsible for ensuring that ASPR has the systems, logistical support, and procedures necessary to coordinate the Department's operational response to acts of terrorism and other public health and medical threats and emergencies. OPEO maintains a regional planning and response coordination capability, and has operational responsibility for HHS functions related to the National Disaster Medical Systems (NDMS).

OPEO acts as the primary operational liaison to emergency response entities within HHS:

- the U.S. Food and Drug Administration,

- HRSA,

- SAMHSA,

- Centers for Disease Control and Prevention,

OPEO also acts as the primary operational liaison to emergency response entities outside HHS:

- HDS,

- United States Department of Veterans Affairs,

- United States Department of Defense),

- and the public..[10]

2.17.4   References

[1] "National Disaster Medical System".   HHS. Retrieved September 11, 2012.

[2] "DMAT". National Disaster Medical System. Retrieved September 11, 2012.

[3] "DMORT". National Disaster Medical System. Retrieved September 7, 2006.

[4] "VMAT". National Disaster Medical System. Retrieved September 7, 2006.

[5] "FCC". National Disaster Medical System. Retrieved September 7, 2006.

[6] "NPRT". National Disaster Medical System. Retrieved September 7, 2006.

[7] "IMSuRT". Mass General Hospital. Retrieved January 12, 2009.

[8] "NNRT". National Disaster Medical System. Retrieved September 7, 2006.

[9] http://www.phe.gov/Preparedness/planning/mscc/ handbook/chapter7/Pages/hhsconcept.aspx

[10] "DMAT/NDMS". National Disaster Medical System. Retrieved July 3, 2008.

2.17.5   Other sources

- Knouss RF, "National Disaster Medical System", *Public Health Rep*, 2001;116(suppl 2):49–52.

2.17.6   External links

- "National Disaster Medical System".   Retrieved September 11, 2012.

- National Medical Response Team - Central

- OH-1 Disaster Medical Assistance Team - Toledo, Ohio

- RI-1 Disaster Medical Assistance Team - Rhode Island

- CA-11 Disaster Medical Assistance Team - Sacramento, California

2.18   National Incident Management System (US)

NIMS redirects here.   For other meanings see Nims.

The **National Incident Management System** (**NIMS**) is a standardized approach to incident management developed by the Department of Homeland Security. The program was established in March of 2004, and is intended to facilitate coordination between all responders (including all levels of government with public, private, and nongovernmental organizations). The system has been revised once, in December 2008. The core training currently includes two courses: (1) IS-700 NIMS, which provides a basic introduction to NIMS, and (2) ICS-100, which includes history, details, and features, along with an introduction to the Incident Command System. Approximately 24 additional courses are available on selected topics. [1]

NIMS standard incident command structures are based on three key organizational systems:

- The Incident Command System (ICS)

- The Multiagency Coordination System

- Public Information Systems

2.18.1   Federal Emergency Management Agency National Integration Center

FEMA's National Integration Center (NIC) has primary responsibility for the maintenance and management of national preparedness doctrine, including:

- Post-Katrina Emergency Management Reform Act of 2006

- National Incident Management System (December 2008)

- Presidential Policy Directive 8: National Preparedness (March 2011)

The NIC relies on its Strategic Resource Group - practitioners and subject matter expertise from state, tribal and local governments, nongovernmental organizations (NGOs) and the private sector – to assist with resource typing definitions.

2.18.2   References

[1] "NATIONAL INCIDENT MANAGEMENT SYSTEM" (PDF). Department of Homeland Security.   September 2011. Retrieved March 29, 2014.

2.19   National Shelter System

The American Red Cross and the Federal Emergency Management Agency (FEMA) together developed the National Shelter System (NSS). Under the National Response Plan now called the National Response Framework auspices, American Red Cross, is the Co-Primary Agency with FEMA responsible for the Mass Care portion of Emergency Support Function #6 - Mass Care, Temporary Housing and Human Services.

The goal of the NSS is to be able to identify the location, managing agency, capacity, current population, needs assessment and other relevant information for all shelters being run during the response to incidents.[1]

2.19.1   References

[1] "National Shelter System". *fema.org*. FEMA. Retrieved 20 June 2015.

2.20   Olney Federal Support Center

The **Olney Federal Support Center** is an underground U.S. government facility owned by the Federal Emergency Management Agency (FEMA) and located in Montgomery County, Maryland.[1] The center occupies 75 acres (30 ha) of property 2.0 miles (3.2 km) west of Laytonsville, Maryland, on the site of a former Nike missile base.[2][3] It is a communications, satellite teleregistration, and data network facility with various functions, both known and unknown to the general public. The facility at one time housed the FEMA Alternate Operations Center (FAOC)[4] as a control center for the National Warning System.[2][5][6] It is also part of the FEMA National Radio System (FNARS), a high-frequency radio network that links FEMA's emergency operations centers.[2]

2.20.1   References

[1] Lew, H.S. (1999). *Supporting Document for Rehabilitation Cost Estimates of FEMA Existing Buildings* (PDF). National Institute of Standards and Technology. Retrieved March 24, 2013.

[2] "Bunkers Beyond the Beltway: The Federal Government Backup System". *The Lay of the Land* (Center for Land Use Interpretation). Spring 2002. Retrieved March 21, 2013.

[3] "Olney O&M Support". General Services Administration. July 25, 2010. Retrieved March 21, 2013.

[4] The Alternate National Warning Center (ANWC) was renamed the FEMA Alternate Operations Center (FAOC) in 1999. See *National Warning System Operations Manual* (PDF). Federal Emergency Management Agency. 2001. p. 1-3. The FAOC may have since relocated to the FEMA Regional Center in Thomasville, Georgia. See Rivera (June 2, 2004). "The Government Underground: FEMA/Military Bases Itch to be Only Game in Town". Portland Independent Media Center.

[5] "Alternate Special Warning Facility: Olney Special Facility". Federation of American Scientists. July 25, 1998. Retrieved March 21, 2013.

[6] Goodwin, Jacob (August 7, 2010). "FEMA Needs Armed Guards at Secretive Federal Support Center in Gaithersburg, MD". *Government Security News*. Retrieved March 21, 2013.

Coordinates: 39°12′33″N 77°06′23″W / 39.2093°N 77.1063°W

2.21   Pets Evacuation and Transportation Standards Act

The **Pets Evacuation and Transportation Standards Act** (**PETS**) was a bi-partisan initiative in the United States House of Representatives to require states seeking Federal Emergency Management Agency (FEMA) assistance to accommodate pets and service animals in their plans for evacuating residents facing disasters.[1] Introduced by Congressmen Tom Lantos (D-California) and Christopher Shays (R-Connecticut) on September 22, 2005, the bill passed the House of Representatives on May 22, 2006 by a margin of 349 to 29.[2] Technically an amendment to the Stafford Act, it was signed into law by President George W. Bush on October 6, 2006.[3] The bill is now Public Law 109-308.[4]

2.21.1   Background

The bill was initiated in the aftermath of Hurricane Katrina when the abandonment of many thousands of pets and other animals brought the matter of animal welfare to national attention.[5] The bill's primary proposer, Tom Lantos, indicated that a press picture of a child being separated from his dog was the bill's catalyst; "The dog was taken away from this little boy, and to watch his face was a singularly revealing and tragic experience. This legislation was born at that moment."[6] On the congressional record for the bill, he explained more fully:

> "The scene from New Orleans of a 9-year-old little boy crying because he was not allowed to take his little white dog Snowball was too much to bear. Personally, I know I wouldn't have been able to leave my little white dog Masko to a fate of almost certain death. As I watched the images of the heartbreaking choices the gulf residents had to make, I was moved to find a way to prevent this from ever happening again."[7]

The Hurricane Katrina animals

Stories of abandoned pets after Katrina filled the media.[8][9] The issue raised questions of class concern, as animal welfare activist noted in the *Washington Post* that some hotels who took in evacuees allowed customers to bring their pets, but those forced to rely on public assistance had no options.[10]

One particular case that garnered widespread attention was that of "Snowball", a small white dog made famous by *Associated Press* reporter Mary Foster's coverage of the evacuation of the New Orleans Louisiana Superdome.[11] The authorities who assisted evacuees onto buses refused to allow pets to board. Foster reported that "Pets were not allowed on the bus, and when a police officer confiscated a little boy's dog, the child cried until he vomited. 'Snowball, snowball,' he cried."[12]

The story of "Snowball" became a centerpiece in fundraising appeals by welfare organizations and various ad-hoc websites were created by people soliciting funds to help locate Snowball and reunite him with the boy.[13] On September 6, 2005 *USA Today* reported that Terry Conger, a veterinarian and information officer for the Incident Command Center that coordinated animal rescue efforts in Louisiana, said state veterinary officers had confirmed that Snowball is safe in a Louisiana shelter and that his owner had been located in Texas.[14] However, it appears the veterinarian officials were mistaken. On September 10, 2005 the Lexington Herald-Leader quoted Dr. Conger as saying that original reports of Snowball's recovery were inaccurate and that "the chances of finding it [Snowball] and returning it to its owner are next to nil".

2.21.2 Opposition

While the bill received wide support, it did have opponents. Two Representatives from the State of Georgia who opposed, Lynn Westmoreland-(R) and Charlie Norwood-(R), announced through spokesmen concerns that the law would unfairly impose federal control over state governance and negatively impact resources from other areas of emergency planning necessary to protect human lives.[6]

2.21.3 See also

- Animal law
- Social effects of Hurricane Katrina

2.21.4 References

[1] Pets Evacuation and Transportation Standards Act House of Representatives website. September 2005. Accessed August 30, 2007.

[2] Shays, Christopher. Animal Welfare: Pets Evacuation and Transportation Standards (PETS) Act House of Representatives website. Accessed August 30, 2007.

[3] President Bush Signs H.R. 3858, the "Pets Evacuation and Transportation Standards Act of 2006" White House (press release). Accessed September 10, 2007.

[4] H.R.3858 Library of Congress. Accessed August 30, 2007.

[5] Nolen, R. Scott. October 15, 2005. Katrina's other victims. The Journal of the American Veterinary Association (JAVMA). Accessed August 30, 2005.

[6] Kemper, Bob. May 23, 2006. Pet-loving Georgians call bill a disaster. Atlanta Journal Constitution. (Reprinted at the House of Representatives site of Congressman Lynn A. Westmoreland). Accessed August 30, 2007.

[7] Pets Evacuation and Transportation Standards Act of 2006 Section 51. United States House of Representatives. September 20, 2006. Accessed August 31, 2007.

[8] see, for example, More and more abandoned pets in New Orleans rescued and Katrina's stranded pets spur massive aid effort.

[9] Scott, Cathy (2008). *Pawprints of Katrina: Pets Saved and Lessons Learned*. Hoboken: Howell Book House. ISBN 978-0-470-22851-7.

[10] Dawn, Karen. September 10, 2005. Best friends need shelter, too Washington Post. Accessed August 30, 2007.

[11] Snowball, Snowball, the little dog who broke the nation's heart! September 7, 2005. PR Leap Business News. Accessed August 30, 2007.

[12] Foster, Mary. September 1, 2005. Superdome Evacuations Enter Second Day Associated Press. Accessed August 30, 2007.

[13] See, for example, Snowball Fund.

[14] Manning, Anita. September 6, 2005. Rescuers scramble to reach animals left in dire straits. USA Today. Accessed August 30, 2007.

2.21.5 Further reading

Irvine, Leslie. 2009. "Filling the Ark: Animal Welfare in Disasters". Philadelphia: Temple University Press. ISBN 978-1-59213-834-0

2.22 Special Flood Hazard Area

A **Special Flood Hazard Area** (SFHA) is an area identified by the United States Federal Emergency Management Agency (FEMA) as an area with a special flood or mudflow, and/or flood related erosion hazard, as shown on a flood hazard boundary map or flood insurance rate map.[1] Areas within the SFHA are designated on the flood insurance rate map as Zone A, AO, A1-A30, AE, A99, AH, AR, AR/A, AR/AE, AR/AH, AR/AO, AR/A1-A30, V1-V30 or V.[2]

Land areas that are at high risk for flooding are called special flood hazard areas (SFHAs), or floodplains. These areas are indicated on flood insurance rate maps (FIRMs).

In high-risk areas, there is at least a 1 in 4 chance of flooding during a 30-year mortgage.

2.22.1 References

[1] United States of America (2002). *Code of Federal Regulations 44, Emergency Management and Assistance.* http://www.gpo.gov/fdsys/pkg/CFR-2002-title44-vol1/pdf/CFR-2002-title44-vol1.pdf: United States Government Printing Office. pp. 313–315.

[2] "Definitions of FEMA Flood Zone Designations". Federal Emergency Management Agency. Retrieved January 25, 2013.

2.23 Stafford Disaster Relief and Emergency Assistance Act

The **Robert T. Stafford Disaster Relief and Emergency Assistance Act** (**Stafford Act**) is a United States federal law designed to bring an orderly and systemic means of federal natural disaster assistance for state and local governments in carrying out their responsibilities to aid citizens. Congress' intention was to encourage states and localities to develop comprehensive disaster preparedness plans, prepare for better intergovernmental coordination in the

Sen. Robert T. Stafford (R, VT)

face of a disaster, encourage the use of insurance coverage, and provide federal assistance programs for losses due to a disaster.[1]

The Stafford Act is a 1988 amended version of the Disaster Relief Act of 1974. It created the system in place today by which a presidential disaster declaration or an emergency declaration triggers financial and physical assistance through the Federal Emergency Management Agency (FEMA). The Act gives FEMA the responsibility for coordinating government-wide relief efforts. The Federal Response Plan implements includes the contributions of 28 federal agencies and non-governmental organizations, such as the American Red Cross. It is named for Sen. Robert Stafford (in Senate 1971– 89), who helped pass the law.

Congress amended it by passing the Disaster Mitigation Act of 2000, and again in 2006 with the Pets Evacuation and Transportation Standards Act.

2.23.1 Titles

Title I: Findings, Declarations and Definitions

Title I provides the intent of Congress to provide continued and orderly assistance from the federal government to state and local governments to relieve hardship and damage that result from disasters. As defined by Title I, an emergency is any instance, or thought that is determined by the President, in which state or local efforts need federal assistance to save lives and protect the health and welfare of the people in a community. A major disaster is defined as any natural catastrophe, fire, flood, or explosion, determined by the president to warrant the additional resources of the federal government to alleviate damages or suffering they cause.[1]

Title II: Disaster Preparedness and Mitigation Assistance

Title II authorizes the President to establish a disaster preparedness program that utilizes the appropriate agencies and gives the President the right to provide technical assistance to states in order to complete a comprehensive plan to prepare against disasters. The President can also administer grants to states to provide funding for the preparation and revitalization of emergency plans.

Title II articulates the necessity of a disaster warning system. This includes the readiness of all appropriate federal agencies to issue warnings to state and local authorities and the disbursement of warnings to the public. This title authorizes the President to make use of either the civil defense communication system or any commercial communications systems that are voluntarily given to the president to issue warnings to the public.[1]

Predisaster hazard mitigation plans were also detailed in Title II. Under this title, the President can establish a program to provide financial assistance to states through the National Predisaster Mitigation Fund. States can then develop a mitigation plan that can lessen the impact of a disaster on the public health, infrastructure, and economy of the community. The President can also establish a federal interagency task force to implement predisaster mitigation plans administered by the federal government. The director of the Federal Emergency Management Agency (FEMA) serves as the chairperson of the task force. Other members of the task force include relevant federal agencies, state and local organizations, and the American Red Cross.[1]

Title III: Major Disaster and Emergency Assistance Administration

Title three explains that upon the declaration of a major disaster or emergency, the President must appoint a federal coordinating officer to help in the affected area. This coordinating officer helps make initial appraisals of the types of relief most needed, establishes field offices, and coordinates the administration of relief among the state, localities, and nonprofits. The President must also form emergency support teams staffed with federal personnel. These support teams are sent to affected areas to help the federal coordi-

nating officer carry out his or her responsibilities. The President also helps with the establishment of regional support teams. Title three also explains the reimbursement process for expenditures by federal agencies under the Act.

The federal government is not liable for any claims based on "the exercise or performance of or the failure to exercise or perform a discretionary function or duty on the part of Federal agency or an employee of the Federal Government in carrying out the provisions of this Act".[1] In general, the expenditure of federal funds for debris clearance, reconstruction, or other emergency assistance which is carried out by contract with private organizations or firms is given to those organizations and firms already residing in or doing business in the affected area.

Title three explains the government's nondiscrimination requirements. The President has the right to issue and alter regulations affecting the guidance of personnel carrying out federal assistance in affected areas. These regulations include provisions for insuring that the distribution of supplies, processing of applications, and other relief activities is accomplished in fair and impartial way without discrimination on the grounds of color, race, nationality, sex, religion, age, disability, economic status, or English proficiency.[1] It also explains that no geographic areas can be precluded from federal assistance by any type of scale based on income or population.

Penalties are set forth in this title. Any person who misuses the funds obtained under the Act may be fined up to one and one-half times the amount that they misused. The Attorney General may also bring a civil action for relief. Any individual who knowingly violates any part of this Act can be subject to a civil penalty of no more than $5,000 per violation.

The last portion of Title three sets forth the requirements of mitigation plans. Each plan developed by a local or tribal government must both describe actions to mitigate hazards and risks identified under the plan and it must establish a strategy to implement those actions. State plans must do four things. The first is to describe the actions to mitigate hazards and risks identified under the plan. Then it must show a way to support the development of a local mitigation plan. The plan must then show how it will provide technical assistance to its local and tribal governments for mitigation plans. Lastly, it must identify and prioritize the mitigation actions that it will support as its resources become available.[1] The President must allow for sufficient public notice and time for public comment before implementation any new or modified policy under this Act that governs the implementation of any public assistance program or that could result in a major reduction of assistance under the public assistance program.

The President shall appoint a Small State and Rural Advocate whose main responsibility is to ensure the fair treatment of small states and rural communities in the provision of assistance under the Act. The advocate may also help small states prepare requests for emergency declarations.

Title IV: Major Disaster Assistance Programs

The procedures for declaring a major disaster are to be made by the governor of the state. When a disaster occurs, the governor executes the state's emergency plan. If the Governor then decides that the disaster is of such severity that the state and affected local governments cannot possibly handle the effects of the disaster, the Governor will make a request to the President explaining the amount of resources they currently have available and commit to the cost-sharing requirements in the Stafford Act. The President can then declare a major disaster or emergency in the affected area.

Title IV sets out the authority of the President during major disasters or emergencies. The president has many powers under this act. These powers include, but are limited to: directing any federal agency to help the affected area (including precautionary evacuations), coordinating all disaster relief assistance, providing technical and advisory assistance (issuing warnings, providing for the public health and safety, and participating in recovery activities), distributing medicine, food and other supplies, and providing accelerating federal assistance when the President deems it necessary. Lastly, the President can also provide any emergency communications or public transportation that an affected location might need. The federal share of these types of assistance is no less than 75 percent of the eligible costs.[1] The President has the ability to contribute up to 75 percent of the cost of any state or local hazard mitigation effort that is deemed as cost-effective and substantially reducing the risk of a major disaster.

U.S. Navy sailors stationed at Naval Air Station Pensacola load supplies on a UH-3H helicopter before it is transported to New Orleans to aid in disaster relief efforts for the Hurricane Katrina victims.

During a major disaster the Governor may request that the President direct the Secretary of Defense to use the resources of the United States Department of Defense for the purposes of any emergency work. This work is only allowed to be carried out for 10 days. Emergency work is defined as "clearance and removal of debris and wreckage and temporary restoration of essential public facilities and services".[1] Title IV also provides a framework for many essential governmental functions during an emergency including legal services, relocation assistance, distribution of food coupons and unemployment assistance.

If, during an emergency, a local government has lost such a substantial amount of revenues that they cannot perform essential government responsibilities, the President is authorized to provide Community Disaster Loans. The loan amounts are based on need and cannot exceed either (1) 25 percent of the annual operating budget of that local government for the fiscal year in which the disaster occurs and not exceeding $5,000,000, or (2) if the loss of tax and other revenues of the local government as a result of the disaster is at least 75 perccent of the annual operating budget of the local government for that fiscal year, 50 percent of the annual operating budget of that local government for the fiscal year in which the disaster occurs, not exceeding $5,000,000.[1]

The federal government will not have the authority to impede the access of an essential service provider to an area impacted by a major disaster. A major service provider is defined as either: a telecommunications service, electrical services, natural gas, water and sewer services, or, is a municipal entity, nonprofit entity, or private entity that is responding to the disaster.[1]

Types of housing assistance are identified under this title. The President can provide financial assistance to be used for individuals wishing to rent alternate housing during a time of emergency. The President may also provide temporary housing units directly to the displaced citizens affected by a major disaster. This type of assistance ends after the 18-month period beginning on the date the President declares the major disaster. The President does have the authority to extend the period if he deems it necessary. The President may also provide funds for the repair or replacement of owner-occupied housing damaged by a major disaster. The federal share of the costs eligible for housing assistance is 100 percent.[1]

Title V: Emergency Assistance Programs

Title V explains the process a state must follow to request that the President declare an emergency. Every request for the President to declare an emergency must come from the governor of the state. In order for a request to be made, the Governor must deem that the situation is beyond the potential for the state to manage. To do this, the Governor must begin execution of the state's emergency plan and detail the types and amount of federal aid that will be required. Upon receiving this information the President can then decide if the situation qualifies as an emergency. The President does have the authority to declare an emergency without the Governor's request if the President determines that the emergency falls within the primary responsibility of the United States exclusive or preeminent responsibility as governed by the United States Constitution or laws.[1]

The specific abilities of the President are also explained in this Title. The President can direct any federal agency to use its resources to aid the state or local government in emergency assistance efforts. He also has the responsibility to coordinate all disaster relief assistance and assist with the distribution of food, medicine and other vital supplies to the affected public. The President can provide assistance with debris removal and provide any needed emergency assistance. This Title also gives the President the authority to provide accelerated federal assistance when it has not yet been requested.

The federal share of the costs of such efforts is to be no less than 75 percent of the eligible costs.[1] Total assistance under this Act for one emergency is to be limited to no more than $5 million, except when the President determines additional funds are needed. If additional funds are needed, the President must report to Congress on the extent of the additional need.[1]

Title VI: Emergency Preparedness

Title six explains the measures that have to be undertaken to prepare for anticipated hazards including creating operational plans, recruiting and training personnel, conducting research, stockpiling necessary materials and supplies, creating suitable warning systems, and constructing shelters. During a hazard, governments are expected to evacuate personnel to shelter areas, control traffic and panic, and control use of civil communications. After a hazard has occurred, governments must provide services such as fire fighting, rescue, emergency medical, health and sanitation. They must also remove debris and repair or restore essential facilities.

Title six also sets out the authority and responsibilities of the director of FEMA. The director may prepare and direct federal plans and programs for U.S. emergency preparedness. The director should also delegate emergency responsibilities to federal agencies and state and local governments. Conducting research and training is another responsibility of the director of FEMA. Research should address issues such as shelter design, effective design of facilities and the standardization of those designs, and plans that acknowledge the needs of individuals with pets and service animals

during an emergency.[1] Training should be provided for emergency preparedness officials and other organizations who participate in emergency situations.

FEMA - 13132 - Photograph by Bill Koplitz taken on April 5, 2005 in District of Columbia

One responsibility of the FEMA director is to oversee the development and follow through of emergency preparedness compacts, otherwise known as Emergency Management Assistance Compacts (EMACs). "The Emergency Management Assistance Compact (EMAC) is an interstate mutual aid agreement that was developed out of the need to assist and coordinate resources across states in the event of a disaster situation."[2] These compacts strive to deliver materials and services quickly to affected areas during an emergency. These plans must be submitted to the Senate and House of Representatives.[1]

The FEMA director has the ability to give financial contributions to the states for emergency preparedness purposes. These purposes typically include construction, leasing, and renovating of materials and facilities.[1] The amount contributed by the director must be equally matched by the state from any source it finds is consistent with its laws.[1] Any contribution given to a state for shelters or other protective facilities is determined by taking the total amount of funds available to the FEMA director for these facilities in a fiscal year and apportioning it among the states in a ratio. "The ratio which the urban population of the critical target areas (as determined by the Director) in each state, at the time of the determination, bears to the total urban population of the critical target areas of all of the States."[1] The states must also equally match these funds. If they cannot, the director may reallocate the funds to another state. The director must also report to Congress at least once a year regarding all the financial contributions made for emergency preparedness.

Title six then explains the requirements for an emergency preparedness plan. The plan must be in effect in all political subdivisions of the state. It must also be mandatory and supervised by a single state agency. The plan must make known that the state must share financial responsibility with the federal government from any source it has determined is consistent with its state laws. It must also provide for the creation of a state and local emergency preparedness plan and the employment of a full-time emergency preparedness director or deputy director by the state. An emergency preparedness plan must also make available to the director of FEMA and the Comptroller General any records, books, or papers necessary to conduct an audit. Lastly, a plan must include a way to provide emergency preparedness information to the public (included limited English speakers and those with disabilities) in an organized manner.[1]

The last portion of Title six addresses security regulations. No FEMA employee is allowed to be in a position of critical importance, as defined by the director of FEMA, until a full field investigation of the employee is completed. Every federal employee of FEMA acting under the authority of Title Six, except those in the United Kingdom or Canada, must complete a loyalty oath as follows:

"I_____, do solemnly swear (or affirm) that I will support and defend the Constitution of the United States against all enemies, foreign and domestic; that I will bear true faith and allegiance to the same; that I take this obligation freely, without any mental reservation or purpose of evasion; and that I will well and faithfully discharge the duties upon which I am about to enter.

"And I do further swear (or affirm) that I do not advocate, nor am I a member or an affiliate of any organization, group, or combination of persons that advocates the overthrow of the Government of the United States by force or violence; and that during such time as I am a member of _____ (name of emergency preparedness organization), I will not advocate nor become a member or an affiliate of any organization, group, or combination of persons that advocates the overthrow of the Government of the United States by force or violence."[1]

Title VII: Miscellaneous

Title VII gives the President the authority to determine any rule or regulation that may be necessary to carry out the powers that he is given in the Act. This can be either through a federal agency, or any other means the President sees fit.

Payment deadlines were also established under this Title. Payment of any approved assistance is to be distributed within 60 days of the approval.[1] Any donation, bequest, or gift received under the subsection is to be deposited into a separate fund on the books of the United States Department of the Treasury.[1] Disaster grant closeout procedures under this Title explain that there should be no administrative action in an attempt to recover any payments made to state or local governments for emergency assistance under the Act until three years after the final expenditure report

has been transmitted for that emergency.[1]

Firearm policies prohibit the confiscation of firearms for any reason other than failure to comply with federal law or as evidence in an investigation. It also prohibits the forced registration of a firearm for which registration is not required any federal, state, or local law.[1] The Title also lays out the rights and legal framework for citizens who feel their gun rights have been violated during a time of emergency.

2.23.2   Criticism

There are many criticisms of the Stafford Act. The Institute for Southern Studies has stated that the Act needs to give greater latitude to FEMA on how it responds to disasters that are extraordinarily devastating such as Hurricane Katrina.[3] This is especially true for FEMA's ability to provide financial assistance in the form of grants to states and localities suffering after such a disaster. The Institute for Southern Studies has also noted the red tape that has been associated with the Stafford Act in the Hurricane Katrina recovery efforts. In an article for *Frontline*, many others agreed that the process of handing out aid was hindered by bureaucratic red tape.[4] This leads to a rather slow response from Washington to diagnose and resolve issues with recovery efforts.

Another criticism levied by the authors of the *Frontline* article included the provision in the Stafford Act that requires buildings that are destroyed to be rebuilt the same way that they were standing before the disaster occurred. For example, if a 50-year-old hospital was destroyed during a disaster, the Stafford Act would require the building to be constructed exactly how it was without any updates to the building.

Other criticisms of the Stafford Act focus on human rights issues that are present during emergencies and recovery efforts. The Stafford Act does not require that the federal government ensure displaced persons have the ability to participate in governmental decisions that affect the recovery efforts.[5] This includes not only access to public forums about recovery planning and management, but the Stafford Act also does not address voting rights or civic participation issues for those who are displaced during a disaster.[5]

Many people argue that while the Stafford Act allows the government to step in and provide housing and medical assistance, it does not require it to do so. Any housing, education, or healthcare provided during an emergency and the recovery efforts are provided at the sole discretion of the federal government.[5] Even the rebuilding of medical facilities is discretionary.[5]

While the Stafford Act does give special detail and instruction about the needs of the disabled and animals during an emergency, it does not specify any requirements for children or the elderly. These groups should be given more consideration during an emergency due to extenuating circumstances that could prevent them from following the same emergency protocol as an average adult.[6]

2.23.3 Relevant court cases

- *La Union del Pueblo Entero v. Federal Emergency Management Agency*

- *Davis v. United States*

- *Freeman v. United States*

- *Saint Tammany Parish v. Federal Emergency Management Agency*

- *McCue v. City of New York*

- *Davidson v. Veneman*

- *Hawaii v. Federal Emergency Management Agency*

- *Cougar Business Owners Association v. Washington State*

2.23.4 Proposed amendments

One proposed amendment to the Stafford Disaster Relief and Emergency Assistant Act is the Federal Disaster Assistance Nonprofit Fairness Act of 2013 (H.R. 592), a bill that passed in the U.S. House of Representatives on February 13, 2013, during the 113th United States Congress. The bill would make religious organizations and religious non-profits eligible to receive federal funding for repairs and re-building of their facilities after a major disaster.[7] The bill passed the House by a large margin, but was criticized by opponents for using taxpayer money to help tax-exempt organizations and for violating the principle of separation of church and state.

2.23.5 Pop Culture

In 2015, the Stafford Act was used in an episode of *House of Cards* as a way for President Frank Underwood to fund his signature jobs program, AmericaWorks. In the episode, Underwood used Stafford Act funds under a Title V declaration of emergency for the District of Columbia, citing high unemployment as a disaster in the District. Under Title V of the Act, the president may make an emergency declaration on behalf of an area that is under Federal jurisdiction, which includes Washington, DC. He directed FEMA and other cabinet departments to use the Stafford Funds for jobs programs in the Districts.

2.23.6 References

[1] "Stafford Disaster Relief and Emergency Act". June 2007. Archived from the original on April 1, 2011. Retrieved April 1, 2011.

[2] "Emergency Management Assistance Compact (EMAC) Operations". United States Department of Homeland Security. May 2005. Retrieved March 28, 2011.

[3] Evans, Desiree. "Study on Government Response to Katrina Highlights need for Stafford Act Reform". The Institute for Southern Studies. Archived from the original on April 8, 2011. Retrieved April 5, 2011.

[4] "The Old Man and the Storm". PBS *Frontline*. Archived from the original on April 8, 2011. Retrieved April 5, 2011.

[5] "Protecting the Human Right to Return with Dignity and Justice After Hurricane Katrina". Advocates for Environmental Human Rights. Archived from the original on April 8, 2011. Retrieved April 5, 2011.

[6] "The Legal Landscape for Emergency Management in the United States". New America Foundation. Archived from the original (PDF) on April 8, 2011. Retrieved April 5, 2011.

[7] "H.R. 592 - 113th Congress". United States Congress. Retrieved 15 April 2013.

2.23.7 External links

- Robert T. Stafford Disaster Relief and Emergency Assistance Act (Public Law 93-288) as amended — FEMA website

- CRS Report for Congress — Federal Stafford Act Disaster Assistance: Presidential Declarations, Eligible Activities, and Funding

2.24 United States Continuity of Operations facilities

United States **Continuity of Operations facilities** are spread throughout the country in various locations. Some sites are run by the Federal Emergency Management Agency (FEMA), while others are run by the Department of Defense.

2.24.1 Alphabet Sites

- Site A: Department of Veterans Affairs Readiness Operations Center - Washington, D.C.

- Site B: Department of Veterans Affairs Medical Center - Martinsburg, West Virginia

- Site C: Department of Veterans Affairs Medical Center - Richmond, Virginia

  - Also provides microwave communications relay to Mount Weather and DoD Site C (Fort Ritchie, Maryland)

- Site E: Veterans Administration Medical Center - Bay Pines, Florida

- Site M: National Military Command Center (NMCC) - The Pentagon, Washington, D.C.[1][2]

- Site R: Raven Rock - Sabillasville, Maryland

2.24.2 Site R complex

- Site C: microwave communications relay on top of Quirauk Mountain

- Site Creed: Underground complex on west site of Site R

- Site RT: Raven Rock communications relay

2.24.3 Presidential facilities

- The White House - Washington, D.C.

- United States Naval Observatory

- Camp David

2.24.4 See also

- Continuity of government

- National Program Office (NPO)

- Post-Attack Command and Control System (PACCS)

- National Emergency Airborne Command Post (NEACP)

- National Military Command Center (NMCC)

- Alternate National Military Command Center (AN-MCC)

- Mount Weather Emergency Operations Center

- Warrenton Training Center

2.24.5 References

[1] USAF Historical Division: "The Air Force and the World-wide Military Command and Control System 1961-1965

[2] Arkin, William. (2005) Code Names: Deciphering U.S. Military Plans, Programs and Operations in the 9/11 World. ISBN 1-58642-083-6

2.25 United States Fire Administration

United States Fire Administration seal

The **United States Fire Administration** (**USFA**) is a division of the Federal Emergency Management Agency which in turn is managed by the Department of Homeland Security located in unincorporated Frederick County, Maryland, near Emmitsburg.[1]

2.25.1 History

The U.S. Fire Administration was organized in 1974 in response to the 1973 National Commission on Fire Prevention and Control report *America Burning*. The report stated that a federal agency needed to be organized to help combat the growing problem of fatal fires happening throughout the country. The USFA manages many of the federal programs related to firefighting including systematic collection of statistics relating to fire incidents (National Fire Incident Reporting System), public fire education campaign materials, and information on grants and funding. They also provide a directory of approved, fire-safe hotels, and information on home fire safety.

2.25.2 Programs

The USFA manages the National Emergency Training Center (NETC) in Emmitsburg, Maryland on a campus acquired from Saint Joseph College in March 1979. The NETC comprises the National Fire Academy as well as the Emergency Management Institute. Firefighters and emergency managers from around the United States and the world attend courses at these academies in order to further enhance emergency services in their communities.

2.25.3 Organization

- United States Secretary of Homeland Security
  - Administrator - Federal Emergency Management Agency
    - Administrator
      - Deputy Administrator
      - National Fire Academy Division
      - National Fire Programs Division
      - Support Services Division

2.25.4 See also

- National Fire Incident Reporting System
- National Interagency Fire Center
- Incident command system
- Resource Ordering Status System

2.25.5 References

[1] www.usfa.dhs.gov. Retrieved 2010-06-21. "United States Fire Administration 16825 South Seton Avenue Emmitsburg, MD 21727"

2.25.6 External links

- U.S. Fire Administration

2.26 United States House Transportation Subcommittee on Economic Development, Public Buildings and Emergency Management

The **Subcommittee on Economic Development, Public Buildings and Emergency Management** is a subcommittee within the House Transportation and Infrastructure Committee.

2.26.1 Jurisdiction

The Subcommittee oversees many federal real estate and economic development programs. The real estate management arm of the Subcommittee, for example, oversees the Public Buildings Service, which is responsible for the infrastructure and use of the Capitol Grounds, the Smithsonian Institution, and the John F. Kennedy Center for the Performing Arts. The Subcommittee also manages the Federal Emergency Management Agency (FEMA), and certain aspects of the Department of Homeland Security.

2.26.2 Members, 114th Congress

2.26.3 External links

- Subcommittee website

2.26.4 References

2.27 Waffle House Index

The **Waffle House Index** is an informal metric used by the Federal Emergency Management Agency (FEMA) to determine the impact of a storm and the likely scale of assistance required for disaster recovery. The measure is based on the reputation of the Waffle House restaurant chain for staying open during extreme weather and for reopening quickly, albeit sometimes with a limited menu, after very severe weather events such as tornados or hurricanes. The term was coined by FEMA Administrator Craig Fugate in May 2011, following the 2011 Joplin tornado; the two Waffle House restaurants in Joplin remained open after the EF5 multiple-vortex tornado struck the city on May 22.[1][2] According to Fugate, "If you get there and the Waffle House is closed? That's really bad. That's where you go to work."[3]

The Index has three levels, based on the extent of operations and service at the restaurant following a storm:[3]

- Green: the restaurant is serving a full menu, indicating the restaurant has power and damage is limited.

- Yellow: the restaurant is serving a limited menu, indicating there may be no power or only power from a generator or food supplies may be low.

- Red: the restaurant is closed, indicating severe damage.

Professor Panos Kouvelis of Olin Business School says Waffle House, along with other chains, such as Home Depot, Walmart and Lowe's, which do a significant proportion of their business in the southern US where there is a frequent risk of hurricanes, demonstrates the benefit of good risk management and disaster preparedness. Because the restaurants have a disaster plan and a cut-down menu prepared for times when there is no power or limited supplies, the Waffle House Index rarely reaches the red level.[1][3]

The Waffle House Index sits alongside more formal measures of wind, rainfall and other weather information, such as the Saffir–Simpson Hurricane Scale, which are used to indicate the intensity of a storm.[3]

2.27.1　See also

- Tornado intensity and damage

- Big Mac Index

2.27.2　References

[1] "What Do Waffles Have to Do with Risk Management?". EHS Today. July 6, 2011.

[2] "What the Waffle House Can Teach About Managing Supply Chain Risk". Insurance Journal. July 19, 2011.

[3] "How to Measure a Storm's Fury One Breakfast at a Time". Wall Street Journal. September 1, 2011.

2.27.3　External links

- FEMA Blog: News of the Day (July 7, 2011) – What do Waffle Houses Have to Do with Risk Management?

- Always Open | Georgia Tech Alumni Magazine

- Colbert Report segment on FEMA's Waffle House Index

- Religious allegations http://www.inquisitr.com/2221494/waffle-house-prayer-photo-goes-viral-see-why-some-are-calling-it-offensive/

2.28　World Trade Center Captive Insurance Company

The **World Trade Center Captive Insurance Company** was created by New York City with funding from the U.S. Federal Emergency Management Agency (FEMA) in July 2004, as directed by Public Law 108-7. The law provides up to $1 billion to create an insurance company to cover the risks assumed by the city and its contractors working without commercial insurance coverage, in claims resulting from work done subsequent to the September 11 attacks.[1]

WTC Captive has been criticized by Congressman Jerrold Nadler for spending $103,700,734 on legal fees while paying out only $320,936 in medical claims.

On June 10, 2010, a new settlement was announced giving plaintiffs $712.5 million and reducing payouts to lawyers.[2]

2.28.1　See also

- September 11th Fund

- September 11th Victim Compensation Fund

2.28.2　References

[1] "A Review of the World Trade Center Captive Insurance Company", Federal Emergency Management Agency, OIG-08-21, June 2008

[2] "New Ground Zero Deal Gives Plaintiffs $712.5 Million", By A. G. SULZBERGER and MIREYA NAVARRO, June 10, 2010, New York Times

Chapter 3

Text and image sources, contributors, and licenses

3.1 Text

- **Federal Emergency Management Agency** *Source:* https://en.wikipedia.org/wiki/Federal_Emergency_Management_Agency?oldid= 686910835 *Contributors:* Magnus Manske, Taw, Rmhermen, DavidLevinson, Infrogmation, Hoshie, KAMiKAZOW, Baylink, SeanO, Kingturtle, Alex756, Didup, Saint-Paddy, WhisperToMe, Wik, CBDunkerson, Mackensen, Dbabbitt, Raul654, Jeffq, Aleph4, Altenmann, Postdlf, Texture, Acegikmo1, Rebrane, TPK, Mattflaschen, Ncox, DocWatson42, Mat-C, Tom harrison, Zigger, Michael Devore, Ssd, Rookkey, Mboverload, Shane Lin, SWAdair, Golbez, Neilc, Antandrus, Beland, MisfitToys, TimothyChenAllen, Nickptar, Neutrality, Joyous!, NightMonkey, Discospinster, Rich Farmbrough, Ffirehorse, Cromis, Flapdragon, Czrisher, Huntster, Mwanner, Tom, Jpgordon, Dragoonmac, Sentience, Walkiped, Jolomo, Scuzz138, Numerousfalx, Pearle, Hooperbloob, Musiphil, Alansohn, PaulHanson, Andrewpmk, Xanxz, Ashley Pomeroy, TommyBoy, Cdc, Rebroad, Harej, Drat, Supermjr, Versageek, Zootm, Seth Goldin, Nightstallion, Richard Weil, Mahanga, Crosbiesmith, Paradiver, OwenX, Scriberius, Jason Palpatine, Timharwoodx, Uncle G, BillC, Ikescs, Dblevins2, Lincher, Duncan.france, Tabletop, Smmurphy, Hughcharlesparker, Zzyzx11, Carlsmith, Prothonotar, Govus, Marskell, Jack Cox, Magister Mathematicae, Rjwilmsi, Koavf, Rillian, Habap, R.Lange, Mm35173, Keimzelle, Titoxd, Pathaugen, Mirror Vax, DDerby, Ground Zero, T smitts, Ewlyahoocom, Subversive, Tysto, Zhengfu, Benjamin Gatti, Simesa, Gwernol, Awbeal, Ziggythehamster, YurikBot, TexasAndroid, Sceptre, Family Guy Guy, RussBot, Rxnd, Ochlocrat, Epolk, Manop, Member, Tgwitty, Gokulpanicker, Badagnani, Holycharly, Rjensen, Cholmes75, Tough Little Ship, EEMIV, BOT-Superzerocool, CLW, Nailbiter, Pesco, BazookaJoe, StevenL, CQ, Schavira, Toddgee, BrassRat, CoolKatt number 99999, LeonardoRob0t, QmunkE, Emc2, Wikipeditor, Curpsbot-unicodify, Lamat~enwiki, Ultrogothe, Allens, Junglecat, Gurukid, NeilN, Gumpo, Soir, C.levin, William Wallace~enwiki, Sardanaphalus, SmackBot, Unschool, Haymaker, Impaciente, Hydrogen Iodide, MJMyers2~enwiki, Rrius, Ssbohio, Ray4389, Antrophica, Rojomoke, Castellanet, Boris Barowski, Lakhim, Macintosh User, Ohnoitsjamie, Daysleeper47, Chris the speller, Bluebot, Deuxhero, Kaid100, Robth, Mexcellent, Peteg913, TheNewMinistry, Falconwings1982, Andy120290, Joe Ralle, X558, Iapetus, Justacip, Wybot, RJBurkhart, BrotherFlounder, DDima, Theunknown42, Gloriamarie, Cody5, Zahid Abdassabur, Lapaz, Scientizzle, Copeland.James.H, Webbrg, Dumelow, Ckatz, UselessatScrabble, Prosped, MBob, Levineps, G1076, Michaelbusch, Octane, CapitalR, Richard75, Wisher, BHammond1, Tawkerbot2, Eastlaw, Rustavo, Clindberg, Dadip6, WeggeBot, Lilifer29, AndrewHowse, Mrrightguy10, Gogo Dodo, Thewinchester, Silleegyrl, Starionwolf, Mallanox, Rougher07, Thijs!bot, Epbr123, Daa89563, Wikid77, Hieronymus Illinensis, N5iln, Eco84, Tellyaddict, Viralmemesis, Escarbot, AntiVandalBot, QuiteUnusual, Jc3, Superzohar, JimDunning, Spencer, Alphachimpbot, David Shankbone, Indiawilliams, Leuko, MER-C, Ryan4314, GoodDamon, FaerieInGrey, RBBrittain, Farquaadhnchmn, Keithicus, Cola19, Cpl Syx, Patstuart, MartinBot, Grandia01, HubmaN, Penpen~enwiki, Jim.henderson, Roberthb, Vox Rationis, BigrTex, UBeR, Writegeist, Uncle Dick, Paris1127, LordAnubisBOT, McSly, Plasticup, ArmadilloProcess, Olegwiki, Elvis58, Kenneth M Burke, Bhredmatt65, Wikimandia, Andy Marchbanks, Hallebb, Lights, Hotfeba, Station1, Gatotsu911, Philip Trueman, Flyte35, Kiranwashindkar, Malinaccier, Ann Stouter, Someguy1221, Naive rm, Masaqui, Redsox04, Lerdthenerd, Angrymansr, @pple, Fallobfan, SieBot, 4wajzkd02, Fanra, Zachpresnall, Dawn Bard, Jay38932006, Yintan, Mymichelet, Aly89, Ks0stm, Int21h, MASTER178, Biehnb, Richard David Ramsey, Jons63, Xnatedawgx, VanishedUser sdu9aya9fs787sads, Faithlessthewonderboy, MenoBot, Blackskyshining, ClueBot, QueenofBattle, Deanlaw, The Thing That Should Not Be, Mild Bill Hiccup, Foofbun, Wildspell, Tiretabletelephone, Jeanenawhitney, Excirial, Jusdafax, Agmonaco, Niteshift36, Mindlessjapican, Jandrews23jandrews23, Halgin, Belchfire, Danjres, Little Mountain 5, Alexius08, Jacebel, Thatguyflint, Thebestofall007, Addbot, Proofreader77, Tcncv, TutterMouse, Matt.Chojecki, MagnusA.Bot, Nathrael, Download, Favonian, Relieffan, Dayewalker, BrianKnez, Luckas Blade, Swarm, Legobot, Luckas-bot, Yobot, Ptbotgourou, Fraggle81, Legobot II, Librsh, AnomieBOT, Bsimmons666, Jim1138, Galoubet, 9258fahsflkh917fas, Mmeytin, TParis, Citation bot, FreeRangeFrog, Nasnema, Inferno, Lord of Penguins, GrouchoBot, Theslider09, MerlLinkBot, Prezbo, Cekli829, Green Cardamom, Jdiaz532, FrescoBot, Surv1v4l1st, Tobby72, Latviyaa, Cs32en, Gire 3pich2005, Sakha Sire, Sunnyday63, I dream of horses, Calmer Waters, Mehrunes Dagon, Mercy11, Dinamik-bot, Vrenator, Trisssz1011, GGT, PleaseStand, RjwilmsiBot, Stefnoem4, EmausBot, John of Reading, WikitanvirBot, Ajraddatz, RA0808, Rockerman777, Wikipelli, Sheeana, Illegitimate Barrister, Josve05a, Bollyjeff, Lateg, The Nut, Anonymous31415, H3llBot, Unreal7, Donner60, Brycehughes, Bootcamp123456789, ClueBot NG, Widr, BG19bot, Sprinting faster, Slapdapope, Compfreak7, EmadIV, VacationLanegrp, Dr. Coal, Cadeslim, Shaun, BattyBot, Hghyux, Breaker 355, Cyberbot II, Nicolewashere, Mediran, Ninjamaster1389, Dexbot, Mysterious Whisper, Mogism, Darrenpeachey, ComfyKem, UrkelforPresident, Epicgenius, Vanamonde93, BreakfastJr, Ldemartino, Heatedfrosr, Originallygeneric, Kathrynellens, Nhj78992, HLS

Student 1241, Guysittingathome, Alfred917, Monkbot, Sandersedgar, Crystallizedcarbon, Catfish33, Emcsherr, Dabom321, Huskers110110, Lordrocko990126, Brecapla000, Jbirds44, Whoshoosier and Anonymous: 499

- **Public health emergency (United States)** *Source:* https://en.wikipedia.org/wiki/Public_health_emergency_(United_States)?oldid=687889333 *Contributors:* The Anome, Welsh, Mike Serfas, SmackBot, Avalon01, PamD, JamesAM, JustAGal, Postcard Cathy, The Anomebot2, Funandtrvl, WereSpielChequers, Mild Bill Hiccup, Yobot, Jburlinson, Snakespeaker, I dream of horses, RjwilmsiBot, John of Reading, Austritch and Anonymous: 4

- **Public Health Emergency Preparedness** *Source:* https://en.wikipedia.org/wiki/Public_Health_Emergency_Preparedness?oldid=686928973 *Contributors:* Matt Crypto, Quadell, Ian Pitchford, SmackBot, Colonies Chris, Iridescent, Canby, Numberman1, Int21h, Shhonn22, Full-date unlinking bot, We hope, Geofferybard, GabeIglesia, ArthurJomasSmith, Brentcopper and Anonymous: 5

- **Community emergency response team** *Source:* https://en.wikipedia.org/wiki/Community_emergency_response_team?oldid=684484002 *Contributors:* Derek Ross, Andre Engels, Ray Van De Walker, SimonP, Ellmist, Ram-Man, Zanimum, Cameron Dewe, DJ Clayworth, Secretlondon, Styxmark, Jmarkantes, Obli, Ssd, Mboverload, Tjwood, Deleting Unnecessary Words, Petershank, Mwanner, CDN99, Evolauxia, Cmdrjameson, Guidod, Charonn0, Vanished user 19794758563875, Paullaw, Espoo, PaulHanson, Goldom, Geraldshields11, Firsfron, Graham87, Vegaswikian, ElKevbo, Mirror Vax, Howcheng, Brandon, Moe Epsilon, Designmotif, SmackBot, Septegram, Pzavon, Betacommand, Kharker, Thumperward, Htra0497, INHUMANITY, Thisisbossi, Luno.org, Soap, Emanz, Cashcraft, Certdmill, JoeBot, Jwalte04, T.O. Rainy Day, Apkinesis II, Neoking, NelsonJacobsen, CmdrObot, Pblum, Dragomiloff, K7aay, DuncanHill, Desertsky85451, JaGa, Calltech, J.delanoy, Saberflash, Parradoxx, Stephendonnelly88, Ushero, Calmariner, Minesweeper.007, Mikeboatlake, InfoQuest, ExtraDry, AgentAnderson, AHMartin, PbBot, Rebelyell2006, Konaallan, Dabomb87, ImageRemovalBot, Certhaltomcity, Joelbarrett, Icannotlivewithoutbooks, Callinus, XLinkBot, Dthomsen8, Pecanpieeater, Billoneil, Lightbot, Legobot, Legobot II, Dregas, Amirobot, J appleseed2, Smallman12q, Thehelpfulbot, Warlink, Crusoe8181, Dolovis, Wpwx694, TheJJJunk, Bostonguy2000, Et0048, Robevans123 and Anonymous: 78

- **CRP-2B** *Source:* https://en.wikipedia.org/wiki/CRP-2B?oldid=633179463 *Contributors:* Uncle G, Rjwilmsi, Malcolma, SmackBot, Addbot, SporkBot and Helpful Pixie Bot

- **Disaster medical assistance team** *Source:* https://en.wikipedia.org/wiki/Disaster_medical_assistance_team?oldid=683075104 *Contributors:* Mwanner, PaulHanson, Scriberius, Mandarax, BD2412, Search4Lancer, Rjwilmsi, Mirror Vax, Ground Zero, Jmorgan, Epolk, Tony1, SmackBot, Ze miguel, Jimacosta, Kc5fm, Prosped, Cydebot, Dawnseeker2000, Peck10, PFucci, Trauma tom, Wb4lai, Responder, Rvrsardog, Flatlyn, 🇮🇳🇮🇳🇮🇳🇮🇳, John of Reading, Kevnalu, Racerx11, H3llBot, Frietjes, Neøn, Ffeld, Khazar2, Cerabot~enwiki, JakeWi and Anonymous: 25

- **Disaster Mortuary Operational Response Team** *Source:* https://en.wikipedia.org/wiki/Disaster_Mortuary_Operational_Response_Team?oldid=517390735 *Contributors:* PaulinSaudi, Rich Farmbrough, Giraffedata, Rjwilmsi, Epolk, RobDe68, WODUP, Bullsswimming, Yobot, FrescoBot, H3llBot, Rcornett, Depcomm and Anonymous: 3

- **Emergency management** *Source:* https://en.wikipedia.org/wiki/Emergency_management?oldid=688684327 *Contributors:* Ray Van De Walker, SimonP, Edward, Mac, Ronz, Zeizmic, Andrewman327, Snickerdo, Postdlf, Alan Liefting, Jrquinlisk, Leonard G., Rchandra, Bobblewik, Stevietheman, Chowbok, Andycjp, Beland, Plasma east, Glogger, Shiftchange, Discospinster, Vsmith, Psyco path industries, Neko-chan, Violetriga, RJHall, Mwanner, Reinyday, John Vandenberg, .:Ajvol:., Nesnad, Pearle, Paullaw, Walter Görlitz, Arthena, DreamGuy, Velella, Shoefly, Sfacets, Vers ageek, SteinbDJ, Firsfron, Woohookitty, Pol098, MONGO, Haunti, Mandarax, Graham87, Sherpa~enwiki, Baeksu, Trlovejoy, Bruce1ee, Ligulem, The wub, Fred Bradstadt, Yamamoto Ichiro, Ground Zero, Cherubino, Cmadler, Russavia, CJLL Wright, Bgwhite, Kjlewis, The Rambling Man, Wavelength, RussBot, Rxnd, Backburner001, Me and, Bhny, Chris Capoccia, Gaius Cornelius, CambridgeBayWeather, Anomalocaris, NawlinWiki, DragonHawk, Aeusoes1, Grafen, Neum, Mkill, Blowdart, Zarboki, Alpha 4615, Nick123, Ketanof92, Zzuuzz, Open2universe, Gregzeng, Ninly, Nikkimaria, Arthur Rubin, Owain.davies, Katieh5584, Mebden, NeilN, Exit2DOS2000, Seifried~enwiki, SmackBot, McGeddon, Sn00kie, Mcmillen76, Lujc~enwiki, Kintetsubuffalo, Direktorxxx, Septegram, Pzavon, Gilliam, AstareGod, Chris the speller, TimBentley, J. Spencer, Thekaleb, Konstable, Badger151, Can't sleep, clown will eat me, Egsan Bacon, KaiserbBot, JonHarder, Ww2censor, Rrburke, MartinRobinson, Cybercobra, Derek R Bullamore, Wizardman, Kukini, FlyHigh, Will Beback, Acebrock, Harryboyles, Kuru, Breno, Emanz, IronGargoyle, Ckatz, Beetstra, InedibleHulk, Hu12, Ces0007, Ft93110, JoeBot, Beanmedic, Mulder416sBot, Linkspamremover, ChrisCork, CmdrObot, Dycedarg, Neelix, Jim182, Cydebot, Cahk, Pblum, Magresta, Trasel, Smbassfisher, Fair Deal, Anthonyhcole, Corpx, Dancter, Dragomiloff, DumbBOT, Kozuch, Arcayne, Timothy.chapuis@medair.org, Leedeth, Dawnseeker2000, Mrhinard, KrakatoaKatie, AntiVandalBot, Seaphoto, Onestorm, Estiveo, Hamaryns, Ingolfson, Erxnmedia, .alyn.post., Harryzilber, Ekabhishek, MER-C, Zenith69uk, Montie46, Magioladitis, Harburgm, Atlanta06, Nandus, MetsBot, Wcfugate, Jonein, Macboots, Teknomegisto, Symbolt, DerHexer, Esanchez7587, Mschiffler, Flowanda, MartinBot, Magellan762, Homestepsafety, Mschel, R'n'B, EdBever, J.delanoy, Boghog, Parradoxx, 72Dino, Wikip rhyre, Vwpatton, Dispenser, OShea4, Pratikranjandas93, Ryan Postlethwaite, Ushero, DadaNeem, Olegwiki, Spacecoyote94, KylieTastic, Sjwk, Inwind, DASonnenfeld, Shannon12, Marekzp, Tadasana, Afluent Rider, Fredrick day, Oikos-team, LeaveSleaves, PDFbot, David Condrey, Okarabulut, Testmaennchen, D12dimensions, Flor!an, LittleBenW, Logan, Timoluege, Sidharthmadan, HybridBoy, Pyroknight, Bantamus, SieBot, Heb, Liquidtool1188, Zachpresnall, Divemd, Ribisi, Bmaycock, Gpinder, Flyer22 Reborn, Tom Worthington, Nopetro, Iain99, DeknMike, Denisarona, Deminimis, Jons63, Goodshoped35110s, Twinsday, ClueBot, Arabjohn, Mikevkay, Anirudhrocking, Mckenzjo01, Mild Bill Hiccup, Tosaka1, Hs4pratt, Shustov, MindstormsKid, Neoluk, Tisdalepardi, Fuwiaj, Wrobrose, Muhandes, Singhalawap, Dekisugi, Empman, Aleksd, Lffavors, Guidepostservices, Ranjithsutari, TimClicks, Ps07swt, Kmcarter, XLinkBot, Jopparn, Danjres, Mrugen Modi, Addbot, HazardsUS, Haruth, TutterMouse, Matt.Chojecki, Gbradt, Adamsrock, CanadianLinuxUser, MrOllie, Arjuno3, West.andrew.g, Curlupndie, Billoneil, Middayexpress, Rohit max, Yobot, Themfromspace, Fraggle81, KamikazeBot, A.thinus, AnomieBOT, Emrgmgmtca, Galoubet, Userresuuser, Powerzilla, Materialscientist, Sherazade10, Citation bot, Aneah, Patty lynn Brooks, Capricorn42, Chakdeprashant, Nasnema, Coreywalters06, Drewpup99, Anna Frodesiak, Prunesqualer, Alvin Seville, Goodwin-Brent, RyanBEdwards, Shadowjams, Bytbox, A.amitkumar, Skcpublic, FrescoBot, Nageh, Foresightsupply, Kwiki, Pinethicket, I dream of horses, MastiBot, RedCamaro, Bt999uk, Meaghan, Garberino, Felix Schrodinger, MEMCI, Elekhh, Mercy11, Adamheien, Yunshui, Lotje, Animal197, Jeroen12321, January, Raykyogrou0, DARTH SIDIOUS 2, Fzanetta, D820, Onel5969, Mean as custard, Mnemotechnik~enwiki, John of Reading, UCTV, Ncipcdir, Winner 42, Wikipelli, K6ka, AsceticRose, Cerfa, Fæ, JoshuaDozor, Bamyers99, H3llBot, Ashburyintl, L Kensington, Quantumor, Kishee4, Donner60, GermanJoe, GrayFullbuster, 28bot, Xanchester, Darshhi, ClueBot NG, Katiker, MelbourneStar, Lorilorilori, Cornbrea4, Willonthemove, Priyadarshini Shalini, Igottheconch, Monsoon Waves, Brothercanyouspareadime, Widr, Kavita mhatre, Jzander, Helpful Pixie Bot, Dorothyfran, Theoldsparkle, BG19bot, Ramesh Ramaiah, CaryGarcia, 2channeler, Northamerica1000, TillF, Cyberpower678, Alyssa18, Martin Cumberbatch, Mediclawyer, Fergus Ross, Havantshire, Docw1, Forrestjohnson04, Nidashah1, Anbu121, Fylbecatulous, BattyBot, Guanaco55, Bagoto,

Ntemc, Austritch, Mrt3366, Bpsand4307, Khazar2, Taycon07, Ekren, Captabhi, Chris.rider81, CountJoshula, TravellerQLD, Rixinton, EmergencyPrepper, Webclient101, Ukrained2012, Shantanu Natu, Copperchloride, Scubadeb, SFK2, Ltccorreia, Planandprep, Cleanbiz, Faizan, Fy432984, Tentinator, Blythwood, Cable97, Unique bhuwan, DavidLeighEllis, Robert4565, Mandruss, Ginsuloft, Siebert1669, Kahtar, Curlupndie85, Lizia7, Aki9810, Meteor sandwich yum, Masteroftheushi, JaconaFrere, Reynoldsor, Khdrager, Allerenenfer, Corsal92, PurpleMonarch, Christaky, BrettofMoore, Harsh Nehra, Civang, Grisnor, USNOTARY, Disasterology1906, CraigKelley62, Randhawaharnir, Huskers110110, Emily Temple-Wood (NIOSH), Royal Wizard, KasparBot, Ceannlann gorm, Imjain, Khuramhaseeb, Sara Joshep and Anonymous: 530

- **Emergency Management Institute** *Source:* https://en.wikipedia.org/wiki/Emergency_Management_Institute?oldid=686517898 *Contributors:* Tony1, Doncram, SmackBot, Chris the speller, Jllm06, The Anomebot2, Sm8900, Adavidb, Funandtrvl, Hotfeba, Niceguyedc, Niteshift36, Duttler, Folklore1, Yobot, ChrisAV8, Wbm1058, DPL bot, Bohemian Baltimore, Huskers110110 and Anonymous: 5

- **Executive Order 12148** *Source:* https://en.wikipedia.org/wiki/Executive_Order_12148?oldid=526241003 *Contributors:* Bryan Derksen, SeanO, Flockmeal, Neutrality, Appleboy, Cwolfsheep, Scuzz138, Jonathunder, PaulHanson, Paradiver, Rjwilmsi, Mirror Vax, Samuel Blanning, SmackBot, CSWarren, Kbdankbot, J36miles, Elongated shorty and Anonymous: 3

- **FEMA camps conspiracy theory** *Source:* https://en.wikipedia.org/wiki/FEMA_camps_conspiracy_theory?oldid=688287433 *Contributors:* Topbanana, Yobot, FrescoBot and Whoshoosier

- **FEMA Photo Library** *Source:* https://en.wikipedia.org/wiki/FEMA_Photo_Library?oldid=660853836 *Contributors:* PaulHanson, Ground Zero, Canley, MNewnham, Billkoplitz, Nobunaga24, CmdrObot, Trivialist, Mmeytin, Surv1v4l1st and Anonymous: 2

- **FEMA Public Assistance** *Source:* https://en.wikipedia.org/wiki/FEMA_Public_Assistance?oldid=648643673 *Contributors:* Welsh, Chris the speller, Jimmy Pitt, Mhockey, Santryl and Vanessa8

- **FEMA trailer** *Source:* https://en.wikipedia.org/wiki/FEMA_trailer?oldid=681603660 *Contributors:* Infrogmation, Klemen Kocjancic, Rich Farmbrough, Arthena, Titanium Dragon, Woohookitty, Uncle G, AnmaFinotera, Edison, Ground Zero, Winstonho0805, Travis.Thurston, Sceptre, Splash, Gaius Cornelius, Aaron charles, J S Ayer, Katieh5584, SmackBot, Castellanet, Chris the speller, Savidan, Kendrick7, Anlace, Robofish, Coredesat, Jaywubba1887, Alan.ca, Eastlaw, CmdrObot, Ekajati, Cydebot, Ebyabe, Wikid77, Johnherrick, Notmyrealname, MarshBot, Chacor, Mary Mark Ockerbloom, Ejj1955, Inks.LWC, RastaKins, VoABot II, Avicennasis, Ponch's Disco, KylieTastic, Falcon8765, Philhoey, Lightmouse, K2500, ClueBot, Niceguyedc, Star Mississippi, John Nevard, PharaohKatt, Lightbot, AnomieBOT, Surv1v4l1st, LWG, ClueBot NG, Ramaksoud2000, Bmusician, Blake Burba, BattyBot, Xoegki and Anonymous: 24

- **Flood insurance rate map** *Source:* https://en.wikipedia.org/wiki/Flood_insurance_rate_map?oldid=669596653 *Contributors:* Stemonitis, BirgitteSB, Tony1, SmackBot, Hmains, Suicidalhamster, Eastlaw, Alaibot, Itsmejudith, Danielfolsom, Rdavi404, Hugo999, Gueneverey, DAK4Blizzard, QuickieWiki, DRTllbrg, Yobot, GermanJoe, Henry3898383 and Anonymous: 3

- **Forward Challenge** *Source:* https://en.wikipedia.org/wiki/Forward_Challenge?oldid=546443699 *Contributors:* Edward, PaulinSaudi, Rockfang, Erik9bot, H3llBot and Brycehughes

- **HAZUS** *Source:* https://en.wikipedia.org/wiki/HAZUS?oldid=651803491 *Contributors:* Bender235, PaulHanson, Ground Zero, Rxnd, Betacommand, Bluebot, Shahubba, Cydebot, Nakor Wiki, Ibarabi, BG19bot, Philipschneider and Anonymous: 12

- **Main Core** *Source:* https://en.wikipedia.org/wiki/Main_Core?oldid=667091316 *Contributors:* Ubiquity, Katana0182, Bender235, Zachlipton, BDD, Tim!, Gareth E Kegg, Benlisquare, Debivort, Badagnani, Tilman, SmackBot, DouglasCalvert, Carolmooredc, Dp76764, Dillard421, Melcombe, JL-Bot, Coinmanj, Ost316, AnomieBOT, VanishedUser sdu9aya9fasdsopa, Cicero in utero, Operdyne7, Mramz88, Lopifalko, Brycehughes, Phoenixia1177, Leeparrott and Anonymous: 10

- **Mount Weather Emergency Operations Center** *Source:* https://en.wikipedia.org/wiki/Mount_Weather_Emergency_Operations_Center?oldid=676825155 *Contributors:* Drseudo, Minesweeper, Tregoweth, Vzbs34, Skyfaller, DJ Clayworth, Dale Arnett, Seth Ilys, DocWatson42, Beland, RayBirks, Bumm13, Neutrality, Dcandeto, Freakofnurture, JTN, MARQUIS111, Pearle, Dhartung, ProhibitOnions, Before My Ken, Rjwilmsi, Worden, Naraht, JdforresterBot, Mark Sublette, Gareth E Kegg, Irregulargalaxies, SFjarhead, Bleakcomb, Hydrargyrum, Nick C, Takeel, American2, Jogers, Arkon, SmackBot, WillAndrews, Imzadi1979, Gilliam, Westsider, Backspace, Elendil's Heir, Esrever, Kuyku, Nagle, Ckatz, Meco, WilliamJE, Choccy, The Letter J, Jsd, Drmilo, ShelfSkewed, Old Guard, No1lakersfan, Cydebot, Chrisrippel, Dino 214, CurlyKrakow, NE2, Arbogastlw, The Anomebot2, Marty55, GregU, R'n'B, Wikip rhyre, Benscripps, Jon335, ArmadilloProcess, Philip Trueman, Bentley4, Everything counts, Ghostwords, Jasonquick, Baseball Bugs, Lightmouse, Pjotr Morgen, Sitush, McKorn, Ktr101, Niteshift36, 4Russeteer, Thingg, Mhockey, DumZiBoT, XLinkBot, Dthomsen8, Yoko Rasierklinge, Addbot, Broadcasting408, Blaylockjam10, Tassedethe, Lightbot, Krano, NeoBatfreak, Yobot, Nslnsweetie, Ulric1313, Citation bot, LilHelpa, Ruodyssey, RightCowLeftCoast, VS6507, Mrzmanwikimail, Brody Kennen, Jonesey95, Full-date unlinking bot, CaminoEsperanza, Altstikman, Zeplex, Illegitimate Barrister, Havermore, H3llBot, Brycehughes, ClueBot NG, CopperSquare, Lholcombe, Wikipooper69, Mark Arsten, Emerine, Twistedpictures1, Sh spamcatch, BDE1982, Kinglycitrus, Cherrycast and Anonymous: 85

- **National Disaster Medical System** *Source:* https://en.wikipedia.org/wiki/National_Disaster_Medical_System?oldid=669199205 *Contributors:* Michael Devore, Jossi, Hurricane111, PaulHanson, Scriberius, Epolk, SmackBot, Jimacosta, Ichthy9, Cydebot, SpellingBot, Yobot, Flatlyn, H3llBot, PeterFucci, JakeWi and Anonymous: 16

- **National Incident Management System (US)** *Source:* https://en.wikipedia.org/wiki/National_Incident_Management_System_(US)?oldid=603382877 *Contributors:* Delirium, Bearcat, Beland, TimothyChenAllen, Rich Farmbrough, Kwamikagami, Pearle, TheParanoidOne, Anthony Appleyard, Iothiania, Stuartyeates, Mandarax, Mirror Vax, Gaius Cornelius, JDoorjam, Tony1, BusterD, David Jordan, SmackBot, Eric-Albert, CapitalSasha, Shaggorama, Folio1701, Cydebot, KleenupKrew, Scottalter, Jalexanderlp, Hotfeba, Wikisteff, Excirial, Bcouzens, VoteyDisciple, AnomieBOT, The Interior, MerlLinkBot, Full-date unlinking bot, Lotje, Dgareylp, Jjensen79, Jjensenndsu, Micwill, Ryano 98 and Anonymous: 39

- **National Shelter System** *Source:* https://en.wikipedia.org/wiki/National_Shelter_System?oldid=667771787 *Contributors:* Pegship, SmackBot, Neddyseagoon, Shaunenglish, Alaibot, Nam1123, Vrac, ArcSb, MTinsman and Anonymous: 3

- **Olney Federal Support Center** *Source:* https://en.wikipedia.org/wiki/Olney_Federal_Support_Center?oldid=663289982 *Contributors:* Naraht, Coffee, Josve05a, Brycehughes, Emerine, Oriole85 and Anonymous: 2

- **Pets Evacuation and Transportation Standards Act** *Source:* https://en.wikipedia.org/wiki/Pets_Evacuation_and_Transportation_Standards_Act?oldid=671779474 *Contributors:* Docu, Beland, Xmoogle, Ground Zero, Wavelength, Aaron charles, SmackBot, Kanabekobaton, Cydebot, Ginsengbomb, Skier Dude, Moonriddengirl, Euryalus, J496, Kumioko (renamed), Tree Kittens, Bad fudge, AuthorAuthor, Laurinavicius, Fraggle81, MerlLinkBot, Surv1v4l1st, LeslieIrvine, Legalskeptic, Wikignome0530, Rocketrod1960, Helpful Pixie Bot, Michipedian and Anonymous: 6

- **Special Flood Hazard Area** *Source:* https://en.wikipedia.org/wiki/Special_Flood_Hazard_Area?oldid=655515889 *Contributors:* Tony1, PamD, Yobot, Calidum, BG19bot, Heatherawalls, MoLiza1!, Ambient Adventurer, Nabil Kabil and Anonymous: 1

- **Stafford Disaster Relief and Emergency Assistance Act** *Source:* https://en.wikipedia.org/wiki/Stafford_Disaster_Relief_and_Emergency_Assistance_Act?oldid=681581766 *Contributors:* Kingturtle, Timc, PaulHanson, GoldRingChip, Drh, Koavf, Titoxd, Wavelength, Jmchuff, Chris the speller, Eric, Cydebot, Sounil, Swliv, Kumioko (renamed), Duffy2032, Tree Kittens, KevCor360, Lightbot, AnomieBOT, JimVC3, Haeinous, Legalskeptic, Sross (Public Policy), Smile234, HistoricMN44, Barbara4321 and Anonymous: 10

- **United States Continuity of Operations facilities** *Source:* https://en.wikipedia.org/wiki/United_States_Continuity_of_Operations_facilities?oldid=651327384 *Contributors:* Woohookitty, Mais oui!, Tdrss, CMG, Niteshift36, AnomieBOT, Theslider09, LittleWink, Brycehughes, Helpful Pixie Bot and Julietdeltalima

- **United States Fire Administration** *Source:* https://en.wikipedia.org/wiki/United_States_Fire_Administration?oldid=656643556 *Contributors:* WhisperToMe, Neutrality, Giraffedata, Pschemp, Markaci, Scriberius, Koavf, SchuminWeb, Kafziel, RussBot, KJPurscell, D Monack, Daysleeper47, Mfinney~enwiki, Eastlaw, JamesAM, Jatkins, Sm8900, Kiranwashindkar, QueenofBattle, Trivialist, FrescoBot, Dlambe3, Zackmann08, Bohemian Baltimore and Anonymous: 10

- **United States House Transportation Subcommittee on Economic Development, Public Buildings and Emergency Management** *Source:* https://en.wikipedia.org/wiki/United_States_House_Transportation_Subcommittee_on_Economic_Development%2C_Public_Buildings_and_Emergency_Management?oldid=670858582 *Contributors:* GoldRingChip, Canley, SmackBot, Andy120290, Alaibot, JustAGal, Nevermore27, Dcmacnut, Funandtrvl, Glo145, Niceguyedc, Tiller54, Americus55, Bigpoliticsfan and Anonymous: 4

- **Waffle House Index** *Source:* https://en.wikipedia.org/wiki/Waffle_House_Index?oldid=683483454 *Contributors:* TommyBoy, F, Ser Amantio di Nicolao, Yomangani, NoDepositNoReturn, KConWiki, Radlib, DASonnenfeld, Toddst1, ShipFan, Addbot, Benjamin Trovato, ZéroBot, Brandmeister, ClueBot NG, Bo allen22 and Anonymous: 4

- **World Trade Center Captive Insurance Company** *Source:* https://en.wikipedia.org/wiki/World_Trade_Center_Captive_Insurance_Company?oldid=535044362 *Contributors:* Ground Zero, TexasAndroid, SmackBot, Accurizer, Erxnmedia and Supertouch

3.2 Images

- **File:Airport_Trapnsportation_Emergency_Preparedness_Program_Exercise.jpg** *Source:* https://upload.wikimedia.org/wikipedia/commons/0/05/Airport_Trapnsportation_Emergency_Preparedness_Program_Exercise.jpg *License:* Public domain *Contributors:* DOE (US Department of Energy) [1] *Original artist:* DOE

- **File:Ambox_globe_content.svg** *Source:* https://upload.wikimedia.org/wikipedia/commons/b/bd/Ambox_globe_content.svg *License:* Public domain *Contributors:* Own work, using File:Information icon3.svg and File:Earth clip art.svg *Original artist:* penubag

- **File:Ambox_important.svg** *Source:* https://upload.wikimedia.org/wikipedia/commons/b/b4/Ambox_important.svg *License:* Public domain *Contributors:* Own work, based off of Image:Ambox scales.svg *Original artist:* Dsmurat (talk · contribs)

- **File:Appropriations_Act_of_2004.jpg** *Source:* https://upload.wikimedia.org/wikipedia/commons/f/f9/Appropriations_Act_of_2004.jpg *License:* Public domain *Contributors:* ? *Original artist:* ?

- **File:Blue_Unit.jpg** *Source:* https://upload.wikimedia.org/wikipedia/commons/d/da/Blue_Unit.jpg *License:* CC BY-SA 3.0 *Contributors:* Own work *Original artist:* Originallygeneric

- **File:Burlando_Building,_National_Fire_Academy.jpg** *Source:* https://upload.wikimedia.org/wikipedia/commons/4/47/Burlando_Building%2C_National_Fire_Academy.jpg *License:* CC BY-SA 3.0 *Contributors:* Own work by uploader *Original artist:* Acroterion

- **File:CERT.jpg** *Source:* https://upload.wikimedia.org/wikipedia/commons/8/8f/CERT.jpg *License:* Public domain *Contributors:* ? *Original artist:* ?

- **File:CGAUXSPatrol.png** *Source:* https://upload.wikimedia.org/wikipedia/en/4/4e/CGAUXSPatrol.png *License:* CC-BY-3.0 *Contributors:* https://www.flickr.com/photos/20448908@N07/14237092398/in/photostream/ *Original artist:* Jonathan James

- **File:Cert_ecc_(7643874586).jpg** *Source:* https://upload.wikimedia.org/wikipedia/commons/6/65/Cert_ecc_%287643874586%29.jpg *License:* CC BY-SA 2.0 *Contributors:* cert ecc *Original artist:* Arlington County

- **File:Classroom_Response_Kit.JPG** *Source:* https://upload.wikimedia.org/wikipedia/commons/f/f1/Classroom_Response_Kit.JPG *License:* CC BY-SA 4.0 *Contributors:* Own work *Original artist:* CraigKelley62

- **File:Commons-logo.svg** *Source:* https://upload.wikimedia.org/wikipedia/en/4/4a/Commons-logo.svg *License:* ? *Contributors:* ? *Original artist:* ?

- **File:Community_Emergency_Response_Team_(US)_Logo.jpg** *Source:* https://upload.wikimedia.org/wikipedia/en/c/c0/Community_Emergency_Response_Team_%28US%29_Logo.jpg *License:* Fair use *Contributors:* https://www.google.com/search?q=Community+Emergency+Response+Team&safe=active&espv=2&source=lnms&tbm=isch&sa=X&ei=s8v3U41IivjJBLHcgrgK&ved=0CAcQ_AUoAg&biw=1821&bih=832&dpr=0.75#facrc=_&imgdii=_&imgrc=6-OuCws5ElldiM%253A%3BdEuaUxkexeoz7M%3Bhttp%253A%252F%252Fwww.whatcomcert.org%252Fassets%252FCERT.jpg%3Bhttp%253A%252F%252Fwww.whatcomcert.org%252F%3B1346%3B795 *Original artist:* ?

- **File:Edit-clear.svg** *Source:* https://upload.wikimedia.org/wikipedia/en/f/f2/Edit-clear.svg *License:* Public domain *Contributors:* The *Tango! Desktop Project*. *Original artist:*
  The people from the Tango! project. And according to the meta-data in the file, specifically: "Andreas Nilsson, and Jakub Steiner (although minimally)."

- **File:Emergency_Responders_(8743399305).jpg** *Source:* https://upload.wikimedia.org/wikipedia/commons/0/0b/Emergency_Responders_%288743399305%29.jpg *License:* Public domain *Contributors:* Emergency Responders *Original artist:* National Institute for Occupational Safety and Health (NIOSH) from USA

- **File:Evacuation_sign.jpg** *Source:* https://upload.wikimedia.org/wikipedia/commons/3/35/Evacuation_sign.jpg *License:* CC BY-SA 4.0 *Contributors:* Own work *Original artist:* CraigKelley62

- **File:FEMA_-_13132_-_Photograph_by_Bill_Koplitz_taken_on_04-05-2005_in_District_of_Columbia.jpg** *Source:* https://upload.wikimedia.org/wikipedia/commons/7/72/FEMA_-_13132_-_Photograph_by_Bill_Koplitz_taken_on_04-05-2005_in_District_of_Columbia.jpg *License:* Public domain *Contributors:* This image is from the FEMA Photo Library. *Original artist:* Bill Koplitz

- **File:FEMA_-_14850_-_Photograph_by_Win_Henderson_taken_on_09-05-2005_in_Louisiana.jpg** *Source:* https://upload.wikimedia.org/wikipedia/commons/0/08/FEMA_-_14850_-_Photograph_by_Win_Henderson_taken_on_09-05-2005_in_Louisiana.jpg *License:* Public domain *Contributors:* This image is from the FEMA Photo Library. *Original artist:* Win Henderson

- **File:FEMA_-_23268_-_Photograph_by_Marvin_Nauman_taken_on_04-03-2006_in_Louisiana.jpg** *Source:* https://upload.wikimedia.org/wikipedia/commons/d/d3/FEMA_-_23268_-_Photograph_by_Marvin_Nauman_taken_on_04-03-2006_in_Louisiana.jpg *License:* Public domain *Contributors:* This image is from the FEMA Photo Library. *Original artist:* Marvin Nauman

- **File:FEMA_-_38464_-_DMAT_team_IOWA-1_assisting_a_resident_in_Texas.jpg** *Source:* https://upload.wikimedia.org/wikipedia/commons/3/3f/FEMA_-_38464_-_DMAT_team_IOWA-1_assisting_a_resident_in_Texas.jpg *License:* Public domain *Contributors:* This image is from the FEMA Photo Library. *Original artist:* Jocelyn Augustino

- **File:FEMA_-_7739_-_Photograph_by_Jocelyn_Augustino_taken_on_03-10-2003_in_Maryland.jpg** *Source:* https://upload.wikimedia.org/wikipedia/commons/2/28/FEMA_-_7739_-_Photograph_by_Jocelyn_Augustino_taken_on_03-10-2003_in_Maryland.jpg *License:* Public domain *Contributors:* This image is from the FEMA Photo Library. *Original artist:* Jocelyn Augustino

- **File:FEMA_-_7740_-_Photograph_by_Jocelyn_Augustino_taken_on_03-10-2003_in_Maryland.jpg** *Source:* https://upload.wikimedia.org/wikipedia/commons/2/2f/FEMA_-_7740_-_Photograph_by_Jocelyn_Augustino_taken_on_03-10-2003_in_Maryland.jpg *License:* Public domain *Contributors:* This image is from the FEMA Photo Library. *Original artist:* Jocelyn Augustino

- **File:FEMA_-_7743_-_Photograph_by_Jocelyn_Augustino_taken_on_03-10-2003_in_Maryland.jpg** *Source:* https://upload.wikimedia.org/wikipedia/commons/f/f6/FEMA_-_7743_-_Photograph_by_Jocelyn_Augustino_taken_on_03-10-2003_in_Maryland.jpg *License:* Public domain *Contributors:* This image is from the FEMA Photo Library. *Original artist:* Jocelyn Augustino

- **File:FEMA_-_7746_-_Photograph_by_Jocelyn_Augustino_taken_on_03-10-2003_in_Maryland.jpg** *Source:* https://upload.wikimedia.org/wikipedia/commons/b/b6/FEMA_-_7746_-_Photograph_by_Jocelyn_Augustino_taken_on_03-10-2003_in_Maryland.jpg *License:* Public domain *Contributors:* This image is from the FEMA Photo Library. *Original artist:* Jocelyn Augustino

- **File:FEMA_-_7756_-_Photograph_by_Jocelyn_Augustino_taken_on_03-10-2003_in_Maryland.jpg** *Source:* https://upload.wikimedia.org/wikipedia/commons/8/84/FEMA_-_7756_-_Photograph_by_Jocelyn_Augustino_taken_on_03-10-2003_in_Maryland.jpg *License:* Public domain *Contributors:* This image is from the FEMA Photo Library. *Original artist:* Jocelyn Augustino

- **File:FEMA_-_7797_-_Photograph_by_Jocelyn_Augustino_taken_on_03-12-2003_in_District_of_Columbia.jpg** *Source:* https://upload.wikimedia.org/wikipedia/commons/f/fa/FEMA_-_7797_-_Photograph_by_Jocelyn_Augustino_taken_on_03-12-2003_in_District_of_Columbia.jpg *License:* Public domain *Contributors:* This image is from the FEMA Photo Library. *Original artist:* Jocelyn Augustino

- **File:FEMA_logo.svg** *Source:* https://upload.wikimedia.org/wikipedia/commons/6/67/FEMA_logo.svg *License:* Public domain *Contributors:* ? *Original artist:* ?

- **File:FEMA_regions.png** *Source:* https://upload.wikimedia.org/wikipedia/commons/f/f0/FEMA_regions.png *License:* Public domain *Contributors:* http://www.fema.gov/about/regions/index.shtm *Original artist:* FEMA

- **File:Fema_trailer_1_Mariel_Carr_Chemical_Heritage_Foundation_Video.jpg** *Source:* https://upload.wikimedia.org/wikipedia/commons/3/38/Fema_trailer_1_Mariel_Carr_Chemical_Heritage_Foundation_Video.jpg *License:* CC BY-SA 3.0 *Contributors:* <a data-x-rel='nofollow' class='external text' href='http://www.chemheritage.org/discover/media/video/culture.aspx#trailers'>*Where Have All the Trailers Gone?*, Mariel Carr, Videographer; Nicholas Shapiro, medical anthropologist and reporter, created for the Chemical Heritage Foundation</a> *Original artist:* Mariel Carr, CHF Videographer

- **File:Fema_trailer_2_Mariel_Carr_Chemical_Heritage_Foundation_Video.jpg** *Source:* https://upload.wikimedia.org/wikipedia/commons/1/12/Fema_trailer_2_Mariel_Carr_Chemical_Heritage_Foundation_Video.jpg *License:* CC BY-SA 3.0 *Contributors:* <a data-x-rel='nofollow' class='external text' href='http://www.chemheritage.org/discover/media/video/culture.aspx#trailers'>*Where Have All the Trailers Gone?*, Mariel Carr, Videographer; Nicholas Shapiro, medical anthropologist and reporter, created for the Chemical Heritage Foundation</a> *Original artist:* Mariel Carr, CHF Videographer

- **File:Firstrespondersexercisetbayontarionov08.JPG** *Source:* https://upload.wikimedia.org/wikipedia/commons/d/d6/Firstrespondersexercisetbayontarionov08.JPG *License:* CC BY-SA 3.0 *Contributors:* Own work *Original artist:* Frmatt

- **File:Flag_of_the_United_States.svg** *Source:* https://upload.wikimedia.org/wikipedia/en/a/a4/Flag_of_the_United_States.svg *License:* PD *Contributors:* ? *Original artist:* ?

- **File:Flag_of_the_United_States_Department_of_Homeland_Security.svg** *Source:* https://upload.wikimedia.org/wikipedia/commons/a/ad/Flag_of_the_United_States_Department_of_Homeland_Security.svg *License:* Public domain *Contributors:* This file was derived from: US Department of Homeland Security Seal.svg
  Original artist: United States Department of Homeland Security

3.3 Content license